TRACK

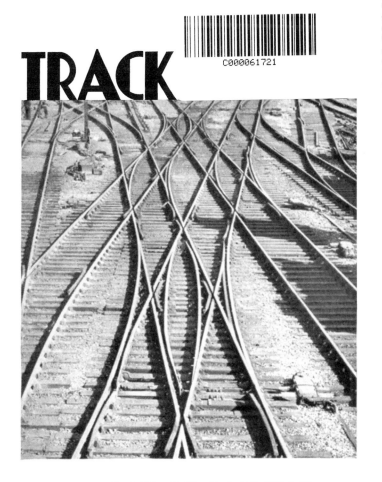

A BOOK OF RAILWAY ENGINEERING
FOR BOYS OF ALL AGES

By W. G. CHAPMAN

(Author of " The 10.30 *Limited," " The ' King ' of Railway Loco-motives," " Cheltenham Flyer," etc.)*

AMBERLEY ·

The main and original part of this book was published in 1935 by the Great Western Railway Company in the 'Boys of All Ages' series. Every effort has been made to produce this facsimile edition as faithfully as possible. For practical reasons the photogravure image of Ponsanooth Viaduct in Cornwall, from the original frontispiece, is now on page 278 in the 'Track Topics Revisited' section. Likewise, the double-page map of the GWR network, has been moved to pages 186-7.

First published 1935 by the GWR.
This edition by Amberley Publishing in 2013

Amberley Publishing, The Hill, Stroud, Gloucestershire, GL5 4EP
www.amberley-books.com

Copyright this edition © John Christopher, 2013

The right of John Christopher to be identified as the Author of this work has been asserted in accordance with the Copyrights, Designs and Patents Act 1988.

All rights reserved. No part of this book may be reprinted or reproduced or utilised in any form or by any electronic, mechanical or other means, now known or hereafter invented, including photocopying and recording, or in any information storage or retrieval system, without the permission in writing from the Publishers.

British Library Cataloguing in Publication Data.
A catalogue record for this book is available from the British Library.

Typesetting and origination by Amberley Publishing. Printed in the UK.

ISBN 978 1 4456 2310 8
E-book ISBN 978 1 4456 2315 3

Introduction to this edition

1935 was a very significant year for the GWR as it marked the company's centenary. Among the 'Big Four' companies the GWR was the only one to have survived the 1923 re-grouping of the railways with its name and territory intact. This gave the GWR a unique sense of continuity. And while its rivals – the LNER, LMS and Southern Railway – all produced various publications of their own, the GWR was able to build upon an established catalogue of railway books. Aside from the usual timetables this had begun with general guide books and the passenger/tourist angle was further exploited through publications such as *Through the Window*, the first volume of which covered the journey from Paddington to Penzance. Issued in 1924, and reprinted in 1927, it depicted the 300 mile route in the form of ribbon maps accompanied by information and illustrations on points of interest. By the late 1920s and into the 1930s other tourist-orientated titles appeared such as *The Cornish Riviera*, *Glorious Devon* and so on, as well as the annual *Holiday Haunts*.

The potential market for books aimed specifically at railway enthusiasts was not neglected. *GWR Names of Engines* appeared from 1911 onwards, and in 1923 *The 10.30 Limited* became the first in a series of 'Railway Books for Boys of All Ages'. A relatively thin

volume of 136 pages, it nonetheless sold in the thousands and was quickly followed by *Caerphilly Castle* (1924), *The 'King' of Railway Locomotives* (1928) and *Cheltenham Flyer* (1934). All of these focused on the glamorous side of the company's activities, the high-profile locomotives, but what of the nitty-gritty work of the engineering department? The first foray in this direction came in 1925 with *Brunel and After*. Subtitled as 'The Romance of the Great Western Railway' it was a good start but concentrated on the history of the line. It wasn't until *Track Topics* in 1935 that the enthusiasts could learn about the minutiae of the civil engineering aspects of the railway; building and maintaining the tracks and the associated viaducts, bridges and tunnels. To my mind this makes it the most interesting of all of the GWR books. As you will see it is divided into a series of talks and, in truth, the conversational style seems somewhat outmoded to the modern reader. But bear with it for *Track Topics* is packed with precious nuggets of useful and insightful information, much of it from a time when the railway was making the difficult transition from its Victorian origins to the fast-changing needs of the twentieth century.

By the 1930s the GWR had established a reputation for the quality of its advertising material, especially the posters designed by well known artists. This is reflected in the unattributed cover artwork for *Track Topics* with its use of flat colour in a bold graphic style. The modern GWR 'button' logo places it firmly in the art deco era and this styling is carried through on the title page. It is a fascinating book, but as you dip into its pages remember that this is far more than just a nostalgia trip. After all, where else will you learn about fly packing and slewing the track, or the difference between a diamond crossover and a double compound?

John Christopher

FOREWORD

SINCE the year 1923 five books have appeared in the Great Western Railway " Boys of All Ages " series of publications. Two of these (with an interval of eleven years between them) have dealt with railway working generally (" The 10.30 Limited " and " Cheltenham Flyer ") ; two others have been concerned with railway locomotives (" Caerphilly Castle " and " The ' King ' of Railway Locomotives ") ; whilst the subject of the remaining volume (" 'Twixt Rail and Sea ") was railway-owned docks.

Now, neither the fastest trains, the most powerful locomotives, nor the most perfect dock system can function without the railway track and its concomitant structures, and it has been represented from time to time that, to round off the series of publications, some account of railway engineering achievements—the track, tunnels, bridges, viaducts, etc.—was wanting.

It is true that in previous volumes an attempt was made very briefly to cover some of the more obvious activities of the Engineering Department, but limitations of space naturally made any such effort entirely inadequate to the relative importance of the subject.

This volume, " Track Topics," sub-titled " A Book of Railway Engineering for Boys of All Ages," is an endeavour to bridge the hiatus. It deals with the civil (as distinct from the mechanical) engineering side of the undertaking, and includes descriptions of some of the

5

FOREWORD

more remarkable engineering features of the Great Western Railway.

In conformity with other volumes in the series, the text is in narrative form (this time as a series of short talks) which, besides being a convenient and intimate medium, is understood to render censure for such heinous offences as sundered infinitives or final prepositions less drastic.

It is hoped that " Track Topics " may both merit and find a place on the bookshelves of boys of all ages alongside " Cheltenham Flyer," which was accorded such a hearty reception last year.

W. G. C.

READING, 1935.

ARRANGEMENT OF TALKS

Typical example of G.W.R. Track.

TALK NUMBER ONE

INTRODUCTORY

On the occasion of your visit to Swindon for the purpose of travelling to London by the world's fastest steam train—" Cheltenham Flyer "—we briefly discussed, among other things, the railway track, and you were then told something about the standard " permanent way " of the Great Western Railway, the arrangements for its maintenance, and the experimental use of steel sleepers.*

Now, I understand, like Oliver Twist, you are asking for " another helping " and this time you would like to be informed about the ways and means of carrying the railway track over and under roads and rivers, through hills and across valleys, and so forth. In short, what you are now anxious to hear more about, is the engineering side of the railway track and its structures.

Perhaps, however, at the outset, I ought to make it clear that what is generally understood by the term " engineering department," as applied to British Railways is the *civil* engineering side of the organisation as distinct from the *mechanical* engineering side.

Such is the paucity of our language that many terms have of unfortunate necessity to be used in many different ways, and of this the word " engine " and its derivatives are examples. The employee termed in this country an engineman, or engine-driver (the " driver " is evidently

---

*See " Cheltenham Flyer."

9

Cornish Riviera Express

a survival from coaching days) is in America known as an " engineer," and it is certainly rather misleading that what is generally known as the engineering department of a British railway is not directly concerned either with the construction or running of railway engines, i.e., locomotives, at all.

Well, your request is a pretty sweeping one, and I think we can best tackle it by a series of talks. I must warn you that these talks will necessarily have to be somewhat patchy and we cannot with advantage maintain any strict order or sequence. Neither can we hope to review the whole of the activities of the Engineering Department of the Great Western Railway, as I am sure you can well appreciate when I tell you that the G.W.R. " geographical " or first track runs to about 3,793 miles or, reduced to single track, about 6,473 miles of running lines and 2,604 miles of sidings, making 9,077 miles in all.

Included in that mileage are about 45,065 switches and 53,390 crossings, 1,675 passenger and goods stations, and 187 tunnels with an aggregate length of 60 miles. There are about 12,000 bridges, to say nothing of cuttings, embankments, and so forth. I don't want to give you a lot of statistics here, or anywhere else, but you may be surprised to learn that the Department which employs from 15,000 to 19,000 men (according to the amount of work in progress) maintains, as well as the railroad itself, about seven million superficial yards of roadway at stations, bridges, etc.

In these talks our theme will be principally the rail-*way* (the track) in its literal sense, and its various structures, although the field of operations of the Engineering Department extends far beyond the track, and the Department

Up from the West

has under its care and maintenance, among other works, a huge dock system which includes breakwaters, piers, jetties, locks with their gates, coaltips, etc., and a large dredging section, and also about 210 miles of canals with aqueducts, bridges, tunnels, locks and wharves.

I think, perhaps, the best plan in setting about our job will be to start with the old Great Western Railway main line from London to the West (Penzance), as originally constructed. Following this route from east to west, we shall encounter some particularly interesting problems of early railway construction in the form of viaducts, embankments, cuttings, bridges, tunnels, sea defence works, etc.

We can then have a look at various phases of permanent-way work of to-day, and consider in turn track components and construction, maintenance, relaying, and the various works for keeping the track level, taking in our stride some of the major structures on other parts of the railway. With such a programme you should become acquainted with a fairly wide and representative selection of works of considerable interest and see how the vast engineering plant of the Great Western Railway of to-day is kept in the van of progress and maintained at that high state of efficiency for which it is so famous.

Let us then, in the first place, go back to the early days of the railway and, with your approval, we will open with a chat about Brunel, the engineering genius whose name is for ever interwoven into the history of the Railway which we are considering. Carlyle tells us that " Great men, taken up in any way, are profitable company," so, to start off, let's get to know something about the great Brunel.

G.W.R. Track—straight as a die

Brunel in characteristic pose

TALK NUMBER TWO

BRUNEL

IT would indeed be difficult to say much about the engineering problems and achievements of the Great Western Railway, without sooner or later (and very much sooner, in fact) referring to the Company's famous first engineer. Before going any farther, therefore, you certainly ought to know something about the man who, more than any other, may be said to have built the Railway.

Isambard Kingdom Brunel, the son of a naturalised French father and English mother, was born on April 9th, 1806, at Portsmouth,* where his father, Sir Marc Isambard Brunel, an engineer of considerable renown, was engaged on behalf of the British Government in the design and erection of machinery for making ships' blocks. Those were the days when mechanical engineering was in its infancy, and the value of the invention may be gained from the fact that, by its aid, work hitherto necessitating the labour of 110 men could be done by the odd ten.

In order to avoid confusion, perhaps, we had better refer to Brunel the younger (with whom we are here concerned) simply as " Brunel," and to his father, as " Sir Marc."

There can be little doubt that Brunel inherited not only a great name in the engineering world, but great natural

---

* The house in which Brunel was born is in Britain Street, Portsea, Portsmouth, where a tablet commemorates the fact.

gifts from Sir Marc and that his education proceeded rapidly under his sire's guidance ; but it would be straining your credulity to assert (as does one encyclopædia) that Sir Marc was assisted by his son in his blockmaking plant, for that work was actually *completed* in the year of Brunel's birth !

We have it on good authority, however, that at the early age of four, Brunel already had a remarkable talent for drawing and that his father had taught him Euclid before he was first sent to school—when eight years old. Perhaps to have mastered the *pons asinorum* before starting school may sound rather an unenviable accomplishment, but Brunel does seem to have displayed really remarkable gifts at a very early age. It has, in fact, been said that " he thought in terms of engineering almost from his cradle."

After two years tuition at Chelsea, he was sent to school at Hove, and in intervals of his classical studies we hear of him making a survey of the town and drawings of the principal buildings.

Brunel was at the College Henri Quatre, Paris, from November, 1820, to August, 1822, studying the language and mathematics, and while there he occupied his leisure in making drawings of engineering works which he sent home for his father's approval.

Although available records emphasise Brunel's extreme devotion as a lad to the pursuit of engineering, there is no reason to assume he was not as full of healthy fun, mischief, and physical activity as other boys. Engineering, however, seems to have had for him the fascination of a favourite hobby as well as a future career.

Without making him out to have been anything in the

nature of a youthful prodigy, young Brunel certainly seems to have known definitely what career he wanted to follow, and prepared for it conscientiously ; more so, I think, than the boy who, asked by a fond uncle what future calling he intended to adopt, said he anticipated becoming an arctic explorer and suggested an allowance of a shilling a day for ices—just to get used to the cold !

Apart from the time spent at Hove and Paris, Brunel seems to have lived almost entirely at home, where he kept in touch with Sir Marc's varied engineering projects, and the advantage gained from this early association was doubtless of the utmost value to him in later life.

When seventeen years of age, Brunel entered his father's office, where one of his first jobs was to assist in work in connection with the Thames Tunnel, of which Sir Marc was engineer. Apparently Brunel was not so fond of the office as of the tunnel works. Construction of the tunnel began in March, 1825, and its whole story, like that of so many tunnels, is one of conflict with natural difficulties. The workings were frequently flooded and on one of these occasions Brunel nearly lost his life. Several times when trouble arose he spent whole nights in the workings, and one record shows he was seven days out of ten in the tunnel, while for nine days his sleep averaged less than four hours !

These facts speak volumes for the dogged perseverance and confidence in himself possessed by Brunel at the very commencement of his professional career, and it is gratifying to learn that such enthusiasm was rewarded by his appointment as resident engineer of the Thames Tunnel

Works in October, 1827, when he was still six months short of his majority.

Probably few young men (or older for that matter) have had higher testimony paid to their devotion to duty than that minuted by the Thames Tunnel Company at its meeting on June 15, 1828. It reads :—

" . . . this Court having heard with great admiration of the intrepid courage and presence of mind displayed by Mr. Isambard Brunel, the Company's resident engineer, when the Thames broke into the Tunnel in the morning of the 12th inst. are desirous to give their public testimony to his calm and energetic endeavours, and to that generous principle which induced him to put his own life in more imminent hazard to save the lives of the men under his immediate care."

One of the extraordinary things about Brunel was the rapidity with which he gained high status as a civil engineer. That is surely definite evidence of remarkable talent, for he passed almost from boyhood to a position of equality with anyone then in the profession.

You may say, perhaps, that all this has little to do with the engineering work of the Great Western Railway, but indirectly, I think, it has *everything* to do with it, for the character which was being forged in young Brunel in those days, was that which, six years later and onwards, was to grapple with and overcome the vast constructional problems of the railway to the West of England. To know something of the man before he accepted the responsibilities of engineer to the Great Western Railway, enables us to appreciate the abilities of the man who built the Railway.

Isambard Kingdom Brunel

It was in 1829, when 23 years of age, that Brunel heard of the project to bridge the River Avon at Bristol. His design was selected in competition with many others and two years later he was appointed engineer for the Clifton Suspension Bridge. It was duly commenced, but, largely owing to lack of funds, work unfortunately had to remain in abeyance for years, and Brunel did not live to see it finished. Doubtless, however, it was this association with Bristol which resulted in his being appointed engineer of a proposed line of railway from Bristol to London at a meeting held at Bristol on March 7th, 1833, when he was still under 27 years of age. Within three months of his appointment he had prepared the preliminary survey and estimates for the railway. It is said that at that time Brunel rarely worked less than twenty hours a day.

TRACK TOPICS

A route via Bradford, Devizes, Pewsey, and Newbury was first inspected, but afterwards that through Bath, Chippenham, Swindon, the Vale of White Horse, and the Thames Valley to Maidenhead, with alternative lines of approach to London, was recommended as the general levels were more favourable, and offered facilities for junctions with Oxford, Cheltenham and Gloucester, By that route, too, Brunel doubtless foresaw the possibilities of extensions into the South Wales coalfields.

A further meeting was called on July 30th, 1833, at the Guildhall, Bristol, when it was resolved to form a company for the establishment of the line of railway of 118 miles (the longest main line hitherto projected) between Bristol and London, and to appoint a body of directors in Bristol who, in conjunction with a similar body in London, would constitute a Board of Management. It was at the first board meeting held in London, on August 19th, 1833, that the title Great Western Railway, which has stood for over a hundred years, was adopted.

A Bill introduced into the House of Commons on March 7th, 1834, just a year after Brunel's appointment, passed its second reading, and went to Committee on April 16th. The Committee stages occupied fifty-four days, and for no fewer than eleven of these, Brunel was under a gruelling examination. He is reported to have acquitted himself valiantly, despite tremendous opposition.

Although that Bill was ultimately thrown out by the Lords, another was promoted the following year, and was read without opposition on March 9th, 1835, committeed, and passed into law on August 31st, 1835, a century ago, in fact.

BRUNEL

Brunel was doubtless very proud of his appointment as engineer of the Great Western Railway, and in his diary at the end of 1835, he wrote : " The railway is now in progress. I am thus engineer to the finest work in England. A handsome salary, on excellent terms with my directors, and all going smoothly."

It was largely due to his enthusiasm and untiring energy that work on the railway was actually commenced at Hanwell the same year, and the first section, London to Maidenhead, was opened in June, 1838. The whole line London to Bristol was open by June 30th, 1841, less than six years from the passing of the Act authorising its construction.

Brunel was a firm believer in speedy transport. " The public," he said, " will always prefer the conveyance which is most nearly perfect, and speed, within reasonable limits, is a material ingredient in the perfection of travelling." You may wonder just what Brunel meant by " reasonable limits," well, I cannot tell you precisely, but he is reported to have said he hoped speeds of 100 miles an hour would be attained on his broad gauge. (We are coming to that in our next talk.)

Now that you know a little about the Great Western Railway's first engineer, we can proceed, but I do not want you to think that work on this Railway and other lines, which later became amalgamated with it, was the full tale of Brunel's railway activities, for he was also engineer of railways in Ireland, Italy, and India.

Brunel will also be remembered for his contribution to steamship construction. *Great Western*, built to Brunel's design and launched in 1838, was then the largest vessel

21

Clifton Suspension Bridge

afloat and the first steam-driven vessel to make regular voyages between this country and America ; while *Great Britain*, the first iron steamship driven by a screw propeller, and that ill-fortuned leviathan *Great Eastern*, were also products of the same fertile brain, and built when it was busy with railway construction.

It is not appreciated perhaps that *Great Eastern* so frequently referred to as a commercial failure, was not only the greatest ship of her time (gross tonnage 18,915), but that her construction definitely established new principles on ship building. Nor is it generally known that *Great Eastern* was not exceeded in size until the building of the White Star Liner " Celtic " in 1901—43 years later !

Brunel was also responsible for dock and harbour works at Bristol, Plymouth, Milford Haven and other places, and his extraordinary versatility was shown in the design for military hospitals used in the Crimean War, and inventions in rifles and large guns.

Perhaps, next to railway construction, however, it is as a bridge builder that the name of Brunel is best known, and his memory is perpetuated at Bristol in the graceful Clifton Suspension Bridge, and in bridges over the Severn, Wye, and Usk. He built some wonderful trestle viaducts, particularly in Cornwall to which (as to much of his other work), we will refer later in these talks, as well as to his masterpiece in bridge work, the Royal Albert Bridge, Saltash.

Brunel was a champion of many schemes, all conceived on grandiose lines, and varied activities ; probably too many, for his life was crowded with work. It has been truthfully said that he was in many respects in advance of his time, but he rose superior to all obstacles and all criticism, and his achievements placed him upon the highest professional plane. Like every genius he certainly had his eccentricities, and it would appear that he was inclined to burden himself with too much work and with much detail which should have been delegated to others.

Unfortunately he did not live to see the completion of some of his finest undertakings, for he died on September 15th, 1859, at the early age of 53. He is buried in Kensal

TRACK TOPICS

Green Cemetery, alongside the Great Western Railway, his best-loved " child."

The Great Western Railway Company has a much treasured possession in the drawing-board used by Brunel. It hangs outside one of the entrances to the Chief Mechanical Engineer's Drawing Offices, at Swindon, and has inspired the following tribute :—

> Within the austere precincts of the hall,
>   Where oft have fallen footsteps of the great,
> A relic of the past adorns the wall,
>   With quiet dignity, and gentle state.
> Not gilded coat-of-arms, nor blatant crest ;
>   Not burnished shield, nor precious-hilted sword ;
> But ranking equal in its own conquest :
>   Isambard Kingdom Brunel's drawing-board.
>
> Historic board that knew a happier day,
>   When that firm hand of genius on you pressed,
> Were you enamoured of the iron-way
>   That blazed a trail of commerce to the West ?
> Were you imbued with vision from that mind
>   In which such mighty projects were evolved ?
> Were you exalted far above your kind,
>   By such a partnership, long since absolved ?
>
> Hang then in modest honour, board,
>   In inspiration may your shadow fall ;
> The high endeavour that your scars record
>   Breathes, in tradition, an insistent call.
> The instruments of genius still shall fill
>   Your erstwhile office : and despite the day,
> Great Western shall grow greater still,
>   While such as you inspire and point the way.

The fame of Brunel is commemorated in his many works : a bronze statue on the Thames Embankment, a window in Westminster Abbey and, quite appropriately, in the " Brunel " medal which is awarded to young railwaymen by the London University (in connection with the London School of Economics) as a token of achievement in railway studies.

Brunel Medal

The original proprietors of the Great Western Railway were not unmindful of Brunel's services, and on January 17th, 1845, they presented him with a service of plate inscribed as under :—

PRESENTED TO

ISAMBARD KINGDOM BRUNEL, ESQ.,

THE ENGINEER-IN-CHIEF OF THE GREAT WESTERN,
THE BRISTOL AND EXETER, THE CHELTENHAM
AND GREAT WESTERN UNION, AND THE
BRISTOL AND GLOUCESTER RAILWAYS,
BY 257 SUBSCRIBERS
TO COMMEMORATE
THE COMPLETION OF THOSE
GREAT NATIONAL WORKS AND TO
RECORD THEIR ADMIRATION OF THE
SCIENCE, SKILL, AND ENERGY MANIFESTED
IN THE DESIGN AND EXECUTION OF THEM, THEIR
GRATITUDE FOR THE ADVANTAGES CONFERRED
ON THEMSELVES AND THE PUBLIC : AND
THEIR ESTEEM FOR THE INTEGRITY
AND WORTH OF HIS PERSONAL
CHARACTER
A.D. 1845.

A section of original Broad Gauge Track

TALK NUMBER THREE

THE BROAD GAUGE AND ITS FATE

AMONG the earliest problems which had to be tackled in building the railway were those regarding methods of constructing the track and its gauge—that is, the distance between the rails.

When work on the railway was commenced, there had been very little experience of railway engineering. Brunel favoured neither the gauge nor the method of construction adopted by Stephenson for the Liverpool and Manchester Railway, which had been opened in 1830. He went about the job of building the Great Western Railway, as he did about most things, according to his own convictions and with a tendency to do everything on a big scale.

All railway engineers of that time were, more or less, pioneers, and being the days of experiments those were also the days of mistakes; and mistakes were made on the Great Western Railway as on other railways.

On the London and Birmingham Railway 152,000 tons of stone blocks had been used as sleepers, and Brunel had originally proposed to use Pennant stone for his railway, but substituted heavy longitudinal timber sleepers. These were thirty feet in length, with cross-ties or "transoms" at intervals of fifteen feet, carried under both lines and laid in pairs at the ends of the longitudinal sleepers, where they were bolted to long beech piles driven into the ground. Brunel also introduced bridge rails of his own

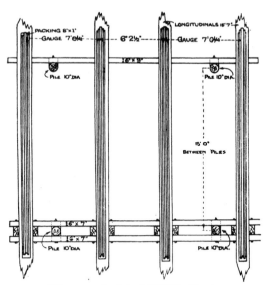

Diagram of original G.W.R. Track

design, another departure from previous practice, which were secured to hardwood planks and fixed, without chairs, direct to the sleepers.

Perhaps I can here anticipate a little by saying that this method of construction, which was adopted for the first section of line from Paddington to Maidenhead Bridge, seems to have caused considerable trouble in various ways, probably due to lack of flexibility in the track and, in the end, the piles used in anchoring the track to the ground were either pulled out or driven down clear of the track.

The Act of 1835, authorising the construction of the railway, contained no stipulation as to the gauge, and this

28

omission was made deliberately at the request of Brunel, who in this matter, as in most other engineering propositions, had views which were very much his own. He was at all times daringly original and unlikely to be influenced overmuch by what others might think on any subject, if opposed to his own considered opinions.

In the matter of the gauge, Brunel was as usual ambitious and, with a foresight not manifested by other railway engineers, he believed that better results could be achieved

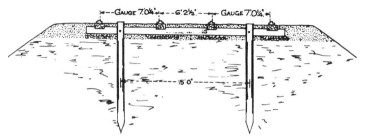

Diagram showing piling

with a greater distance between the rails. This would permit the use of wide, comfortable coaches, slung between wheels of large diameter, thus reducing friction and securing a low centre of gravity, which would make travel at really high speeds, such as had not even been attempted elsewhere, both comfortable and safe.

In his presidential address before the Institute of Civil Engineers the late Sir James C. Inglis, who was himself Chief Engineer, and later became General Manager of the Great Western Railway, said of Brunel, "when the majority of railway engineers and promoters were looking to railways as mere improvements of public roads, on which vehicles were permitted to pass, on payment of tolls, he had a clear perception of their enormous possibilities, in regard to the conveyance of both goods and passengers at high speed."

By the Act of 1835 the railway was to share a London terminus with the London and Birmingham Railway* by means of a junction " in a certain field lying between the Paddington Canal and the Turnpike Road leading from London to Harrow on the western side of the General Cemetery in the Parish or Township of Hammersmith in the said County of Middlesex." The primary scheme is traceable in the direction of the line towards Willesden, which might otherwise have taken a directly westward course.

Negotiations with the London and Birmingham Railway fell through, however, and the directors decided on a London terminus of their own at Paddington. It is reasonable to assume that this change pleased Brunel, for one of the possible objections to his wider gauge would doubtless be the difficulties which would arise at the junction with the narrow gauge (4 ft. 8½ ins.) of the London and Birmingham Railway, and the provisions which would have to be made to meet them.

Brunel had lost no time in reporting to his directors as to the desirability of adopting a wider gauge than that favoured by Stephenson (which was, presumably, a more or less haphazard perpetuation of the width of the old mineral plateways of the time), and by October, 1835, he had obtained sanction to construct the Great Western

---

*Afterwards the London and North Western Railway, and now the London, Midland and Scottish Railway. The lines are only about a third of a mile apart at Kensal Rise.

Section of Brunel's Bridge Rail

Railway to a gauge of 7 feet (actually 7 ft. o¼ ins.), which from that time was known as the " broad gauge."

This is not intended to be a history of the Great Western Railway (for that has been so admirably done by others), but rather a series of talks about some of its engineering problems and achievements and, in order to give you the story of Brunel and his broad gauge, we must now depart from chronological sequence.

During the decade 1835-1845 a good deal of railway construction was undertaken throughout the country and, while the Great Western Railway was gradually extending westwards to Bristol and beyond, by amalgamations and absorptions of broad gauge connecting lines, largely of Brunel's construction, narrow gauge lines were coming into being much faster, and a flood of proposals for more new railways was being launched.

Brunel had come in for a good deal of hostile criticism both on account of his track construction (already described) and his broad gauge, and in order to clear the air other engineers were asked to investigate these matters, with the result that, after careful consideration of their reports, the directors (while agreeing to the abolition of the piles in the track) favoured the retention of Brunel's broad gauge.

At this point, perhaps, it ought to be explained that Brunel was not alone in departing from the narrow gauge, and about a hundred miles of what is now the London and North Eastern Railway had been constructed to a gauge of 5 ft., while in Scotland lines had been laid of 4 ft. 6 ins. and 5 ft. 6 ins. gauges. The Surrey Iron Railway, the first railway for which an Act of Parliament was obtained, was of 4 ft. gauge.

As time went on feeling began to run high on this matter of railway gauge and when lines of narrow (4 ft. $8\frac{1}{2}$ ins.) and broad (7 ft. $0\frac{1}{4}$ ins.) met, as they did at Gloucester, difficulties naturally arose and these brought the matter into the arena of popular controversy. Sides were taken, not only by engineers, financiers, and the press, but by the people, and a spate of pamphlets and press articles appeared from the pens of advocates on both sides.

When the " Battle of the Gauges " was at its height a Royal Commission was appointed to investigate the relative merits of the broad and narrow gauges. There were then 1901 miles of narrow and 274 miles of broad gauge track in the Kingdom, with breaks of gauge at ten points.

The result was perhaps inevitable, for the standard gauge of the British Isles was doubtless settled in 1825, when Stephenson set it for the future of all steam railways. It is interesting to recall, however, that the Commission which sat from August to December, 1845, examined forty-eight witnesses. Thirty-five were in favour of the narrow gauge, whilst eight, all locomotive engineers, advocated a gauge of from 5 ft. to 6 ft., and the remaining four witnesses, who included Brunel and his locomotive superintendent, Mr. Daniel Gooch, were broad gauge champions.

It was Brunel who suggested that the Commissioners should attend a series of engine trials. The narrow gauge party, on various pleas and pretexts, refused anything in the nature of long-distance tests, and it was agreed that trials should be made between London and Didcot (53 miles) and York and Darlington (45 miles) with loads of 60, 70, and 80 tons. These trials brought out the superiority of broad gauge engines and completed the evidence taken by the Commissioners.

In their report of 1846 the Gauge Commissioners said :

" We feel it a duty to observe that the public are mainly indebted for the present rate of speed and the increased accommodation of railway carriages to the genius of Mr. Brunel and the liberality of the Great Western Railway Company."

Apart from all other considerations, a standard gauge was, of course, desirable. We cannot, however, in these days appreciate just how the broad gauge champions viewed the matter when railways were in their infancy. Brunel was broad-minded and far-seeing in many ways, but he seems to have under-estimated the inconvenience which would arise from having more than one railway gauge. It was, of course, far more economical to reduce the " broad " gauge to narrow, than the reverse, and having in view the preponderance of narrow gauge mileage then in existence, the decisions of the Commissioners must have been something in the nature of a foregone conclusion.

Mixed Gauge Track

The Commissioners recommended that 4 ft. 8½ ins. should be declared the gauge for all public railways under construction in Great Britain, and that without the consent of the Legislature, no Company should be permitted to alter the gauge of its railway, and means should be found at points where the two gauges met for narrow gauge vehicles to pass over broad gauge lines.

These recommendations were severely criticised in many quarters, and the Board of Trade could not fully support them, but the " Act for regulating the gauge of railways " of 1846 laid it down that all new railway construction should be of 4 ft. 8½ ins. (henceforth " standard ") gauge unless Parliament expressly sanctioned otherwise. Exceptions were, however, made in respect of certain railways in the West of England and South Wales, all ultimately to become part of the Great Western Railway, and it was laid down that some of the broad gauge lines

34

should provide a third rail (thus forming what was known as " mixed gauge ") so as to admit of narrow gauge vehicles passing over them.

That was the death knell of Brunel's ambitious broad gauge, for although Acts were later obtained for a certain amount of broad gauge construction in Wales, Devonshire, and Cornwall, it gradually became apparent that ultimately the broad gauge would cease to be. Mixed gauge was introduced over various sections of the railway, and, as the advantages of exchange facilities with other railways were becoming more apparent, conversions from broad or mixed gauge to narrow were next undertaken.

Although there were at one time about 270 miles of mixed gauge track on the G.W.R. System, trains of mixed gauge vehicles (i.e., broad and narrow gauges) appear to

Old Signals in use at time of Gauge Conversion

have been worked only to a comparatively small extent,* but presumably this would have been both a practical and an economical arrangement, and it is one which has been adopted satisfactorily on overseas railways of mixed gauge.

In November, 1854, the narrow gauge railways of Shrewsbury and Chester were absorbed. Various other amalgamations were carried out by the Great Western Railway, which in 1861 leased the West Midland Railway for a period of 999 years. The absorption of the West Midland Group (which itself consisted of an amalgamation of 217 miles of track constructed in connection with the Oxford, Worcester, and Wolverhampton Railway) gave the Great Western Railway valuable rights, especially in regard to the route to the North. The lease provided for the laying down of narrow gauge lines between Reading and London, which was effected by putting in a third rail. Owing to the large numbers of narrow gauge vehicles then owned by the Great Western Railway, the third rail was laid to Swindon, Gloucester, and even reached Exeter.

By 1869 it had become apparent that the mixing of the gauge over the broad gauge sections had great drawbacks, especially on the score of expense, and in that year a policy of gradual conversion from broad to narrow gauge was adopted.

The first important conversion consisted in the removal of one rail from the mixed gauge lines north of Oxford, and this was followed by others. In 1869 about $131\frac{3}{4}$ miles of broad gauge were altered to narrow, and in the following decade $648\frac{1}{4}$ miles became narrow gauge, while between

---

*Passenger trains of mixed stock were operated on the Windsor Branch for a time, and goods trains of mixed stock on the West Cornwall Line.

Gauge Conversion operations

1880 and 1891 about 80½ miles were converted. In the year 1892 the remaining 415¼ miles, all between London and Penzance, were dealt with and in all about 1,376 miles of G.W.R. broad gauge track were converted to narrow (or standard) gauge.

It is interesting to note that broad gauge construction went on until as late as 1877 when the St. Ives Branch was opened, although conversions from broad to narrow gauge had then been undertaken on other parts of the G.W.R. system for some years.

The final conversion (and this was only one of several big conversions) was a marvellous engineering achievement, as well as an excellent piece of organisation, and it was carried out with a minimum of inconvenience to the travelling public on the lines west of Exeter, where 168 miles of mixed gauge track were converted to standard gauge in thirty-one hours on May 21st and 22nd, 1892. The most complete arrangements had been made beforehand, and about 5,000 workmen were drafted by special trains from various parts of the system to the territory affected, and accommodated in goods sheds, waiting rooms, and tents. By midnight on May 20th, the last broad gauge vehicle had been moved east of Exeter, to which point mixed gauge had been laid years before.

Dawn on Saturday, May 21st, saw the final work of conversion (which consisted of separating the longitudinal timbers, shortening the transoms, and closing up the timbers again) well in hand from Exeter to Truro, including branches. By a prodigious effort, which called forth a congratulatory telegram from the directors, the whole job was completed by Sunday, May 22nd, and Brunel's gauge was no more.

It had been a heroic contest and *Punch* treated the defeated champion with courtesy and respect, quoting the words, "Good-bye, poor old Broad Gauge, God bless you," which were found chalked on the G.W.R. track. *Punch* also made the end of the broad gauge the subject of one of its famous cartoons by Linley Sambourne (to which suitable verses were supplied), entitled " The Burial of the Broad Gauge," the original of which is at the Company's headquarters at Paddington, as also is a lament, " in loving and regretful memory of the broad gauge," (appropriately written on black-edged paper) by an Eton College boy. You would like to see it ? Well, here is a copy.

There was, indeed, a good deal of genuine sentiment about the passing of the old broad gauge. The last through down train on the old gauge to Penzance left Paddington at 10.15 on the morning of May 20th, drawn by the

This cartoon from PUNCH, *depicting the " Shades of Brunel " at the grave of a broad-gauge locomotive on the occasion of the final gauge conversion in May,* 1892, *was accompanied by verses by Edwin J. Milliken, " The Burial of the Broad Gauge," the last of which read :—*

> Slowly and sadly we laid him down,
> He has filled a great chapter in story ;
> We sang not a dirge, we raised not a stone,
> But we left the Broad Gauge to his glory.

The last Broad Gauge Train passing Sonning on May 20th, 1892

well-known engine " Great Britain," amid sympathetic cheers from a large number of spectators who recognised the significance of the occasion, and similar demonstrations took place at Swindon, Bristol, Exeter, Plymouth and Newton Abbot, where large crowds assembled to give expression to their feelings.

The last broad gauge train to leave Paddington Station was the 5.0 p.m. for Plymouth on May 20th, and the last broad gauge up train to Paddington was the mail which arrived at 5.0 a.m. on May 21st. These two trains met, and stood alongside one another at Teignmouth Station,

when the passengers spontaneously joined hands through the carriage windows of opposite compartments and sang " Auld Lang Syne."

To this day there are those who think that Brunel should have had his way, and that his broad gauge would have simplified many of the problems incidental to railway travel in this twentieth century. It is, perhaps, useless now to speculate as to what could have been done in the way of increased speeds and loads with an addition of 28 inches in the width of the gauge, but it certainly would have made possible carriages and wagons of far greater

41

IN LOVING AND REGRETFUL MEMORY OF
THE BROAD GAUGE

Died May 20th, 1892

On May the twentieth a train
Was speeding o'er the sweet Thames plain
    Upon the Broad Gauge Line.
And from the whistle moans arise
That rend our hearts ; and loud she cries,
    " O, woeful fate is mine ;
Next morn to Swindon shops I go
Where hammers clank and forges glow,
Where I was born scarce four years past,
And yet this trip shall be my last ;
    ' Great Western ' is my name.
I thought to run for many a year
Upon the line whose name I bear
    And win for swiftness fame ;
But now my trav'ling days are o'er
This track shall know me now no more."
She spake, and as her plight I spied
In wrath " Ye Fools and blind " I cried
    " O brutal, bungling, Board
That dare Brunel's great work undo
A greater man than all of you
    And hope thereby your hoard
To heap on high with L.S.D.
May ruin ruthless rout your glee.
Brunel's sad shade I see it stand
With Gooch from Clewer, hand in hand
    And scan the Iron Way,
Not bright and polished as of yore
By thund'ring trains with rush and roar,
    But rusting in decay."

    *   *   *

Next morn in sorrow rose the Sun,
He saw with shame the deed was done,
    Broad Gauge had passed away.

The original written by an Eton College boy is preserved at the
Headquarters of the Great Western Railway.

capacity, whilst locomotive engineers would doubtless have made good use of the extra width available in the construction gauge.

Much has been learned in the century which has elapsed since the Great Western Railway introduced the broad gauge, however, and were Brunel alive to-day he would probably regard with surprise the huge ten- and fourteen-wheeled locomotives and heavy restaurant car and freight trains operating (for example) on the South African Railways with a gauge of only 3 ft. 6 ins., just *half* the width of his " broad " gauge !

The standard gauge has not only rendered possible the through running of trains to all parts of this country, but as the gauge of many of the railways abroad is practically the same as the British, the uniformity of gauge enables a truck of goods to be conveyed (by means of the train ferry across the North Sea) from (say) Bodmin to Bucharest, or to any railway station in Europe (except in Spain, Portugal, and Russia) without transhipment.

Wharncliffe Viaduct

TALK NUMBER FOUR

WHARNCLIFFE VIADUCT—MAIDENHEAD BRIDGE, SONNING CUTTING, ETC.

As we have seen, construction of the railway commenced a century ago at Hanwell, and here at the very outset an engineering problem of no mean dimensions faced Brunel—that of carrying his railway across the wide Brent Valley.

He decided on a viaduct in preference to a continuous embankment, and designed a handsome structure 896 feet in length and 65 feet high, consisting of eight semi-elliptical arches of 70 feet span, springing from piers each composed of twin stone-capped pillars united by a heavy architrave. The viaduct was named after Lord Wharncliffe, as a compliment to the Chairman of the House of Lords Committee, which examined the Bill authorising the railway.

These photographs of Wharncliffe Viaduct will give you a far better idea of that imposing structure than any description of mine, and perhaps you will get a better estimate of its proportions when I tell you that the base area of each of the piers is no less than 252 square feet. The foundations of these piers are carried down, through the gravel, three feet into the London clay. As constructed by Brunel, the viaduct was 30 feet wide, but in the " seventies " it was widened when the lines upon it were quadrupled. Commenced in February, 1836, and completed in the following year, it was quite the most formidable

Wharncliffe Viaduct to-day

single engineering proposition in making the railway between London and the crossing of the Thames at Maidenhead.

The embankments approaching Wharncliffe Viaduct extend for a distance of a mile and a half and attain a maximum height of 63 feet at the west end of the viaduct.

In Wharncliffe Viaduct, we have an example of a structure provided nearly a century ago carrying to-day the heaviest trains at the highest speeds, thanks to the prophetic imagination and engineering skill of its designer.

A short distance west of Wharncliffe Viaduct the railway passes by a bridge over the Uxbridge Road a: a point where it is intersected by the Brentford-Greenford Roads and, as a result, the breadth of span required was considerable. This crossing was contrived by means of an ingenious form of skew bridge, carried on two rows of eight columns each, between the roadway and footpath. Here is an illustration of this unique bridge, which was damaged by fire in 1847, owing to timbers below the permanent way getting alight, and has since been rebuilt.

As you already know, the first section of the line, completed and opened to the public in June, 1838, was from Paddington to Maidenhead Bridge Station (now Taplow), just short of the point at which the railway was to cross the river. The railway between the Brent Valley and Paddington necessitated the provision of a number of bridges and certain road diversions, etc., but as these were chiefly engineering (as distinct from *railway* engineering) problems, we will say no more about them as there are so many other essentially railway problems to discuss.

47

Skew Bridge over Uxbridge Road

It might be added, however, that in order to get ready for the opening of the line from London to Maidenhead, a temporary passenger station was opened on the site of what is now the Paddington Goods Station, and that terminus had to suffice until 1854, when the present station was opened.

Brunel was now confronted with the problem of carrying the railway across the Thames at Maidenhead, at a point where the river is about 100 yards wide. Luckily, at the crossing point, there was a shoal practically in mid-stream and on this he placed the central pier for his bridge.

Unconventional and audacious as ever, Brunel designed a bridge with two of the flattest and largest arches (springing from the central pier) which, up to that time at any rate, had ever been carried out in brick-work. Each of these arches had a span of 128 feet with a rise of $24\frac{1}{2}$ feet above the central pier while, to complete the bridge, there were four semi-circular arches at either end abutting on embankments.

Brunel must have been getting used to adverse criticism by this time, for he certainly had plenty of it. When his

plans for Maidenhead Bridge became known, some of the
" experts " of the period strongly condemned them, stating
that such flat arches could not possibly stand. There was
doubtless joy in the hearts of those experts when, on the
centerings of the bridge being removed, the eastern arch
disclosed a slight distortion.

It was actually a separation of about an inch in the
lowest three courses of bricks, due to the centering being
eased before the cement had time properly to set. The
contractor admitted that he alone was to blame and the
fault was duly remedied. When the necessary amount of
rebuilding was finished, the centres were eased, but by
Brunel's instructions they were to be left in position until
another winter had passed.

Those who were criticising Brunel's work professed to
have scientific knowledge of such and matters solemnly
declared that, when the centres were removed, the arches
must certainly collapse. In the autumn a severe storm

Maidenhead Bridge (from an old print)

Elliptical Arch of Maidenhead Bridge

was experienced and so violent was it that the centres were blown down and, much to the discomfiture of the pundits, Brunel's arches remained standing as they have stood ever since, carrying an increased burden of traffic as the years passed. It was only after the centres had blown down that it transpired that the arches had actually been standing *without* support since the spring of the year !

This confounding of his critics must have pleased Brunel, and it was certainly a feather in his cap. Had he lived he would have been gratified to learn that forty-four years later, when traffic requirements necessitated the widening of Maidenhead Bridge, the new construction was on exactly the same lines as his original design.

It must have been rather disturbing (don't you think ?) for some of those wise people who had been at such pains to prove Brunel's schemes quite impracticable, to find that they were, in due course, successfully accomplished. You see, the worst of saying a thing cannot be done is that you may be interrupted by someone doing it.

MAIDENHEAD BRIDGE

Having crossed the Thames the railway line was soon extended westwards mainly through a long cutting to Twyford, and it was between that station and Reading that another engineering problem of some magnitude was encountered. Brunel had originally intended to take his lines somewhat north of the route adopted and to pass under Holme Park, Sonning, by means of his first railway tunnel, rather more than half a mile in length. He changed his mind, however, and decided on a deep cutting, with a maximum depth of 50 feet, through the high ground met with at this point. This was Sonning Cutting, which at one point passes under the main London-Bath road, carried over the cutting by means of a three-arched bridge—50 feet above the rails.

Power excavating appliances were unknown in those days, and the work had all to be carried out by hand labour —pick, shovel and barrow. As there were 700,000 cubic yards of earth to be moved it was no small job, and at one time 196 horses and 1,220 men were simultaneously employed. Progress was slow at first, but later when locomotives could be brought into use for taking the earth away, as much as 35,000 cubic yards was moved in a week.

Like the rest of the track to Bristol, that through Sonning Cutting was originally for a double line of rails (Up and Down lines) and Brunel's generous provision for his broad gauge was appreciated later, for it made the quadrupling of the track through the cutting in 1893 a much less costly operation than would otherwise have been the case.

When talking of cuttings more generally, we will consider the subjects of slips or settlements in these and also

Early Print of Sonning Cutting

in the sides of embankments and the remedies adopted.
Now we will follow the rails through Sonning Cutting and
we soon come to Reading Station, the largest and most
important non-terminal station on the railway up to that
time.

When the rails reached Reading in 1840, the town was
almost exclusively located on the south side of the line,
and Brunel, to meet this circumstance (which also obtained
at certain other towns) designed a curious one-sided
station in which the Up and Down platforms were at the
east and west ends of the station respectively, and both on
the Down side. It was, in fact, two separate stations
connected by a continuous platform.

This plan of the peculiar lay-out of the lines adopted by
Brunel at Reading Station is well worth a little study, but
in considering it you must remember that in this, as in so
much railway work of the period, the problems were new
with few precedents to draw upon.

SONNING CUTTING

There was certainly one point in favour of this station design, that the passengers had no need to cross the rails, and another that through trains passed the station clear of the platforms. As against these advantages, however, Up trains stopping at the station had not only to cross the Down Line at two points, but also the Down Loop before reaching the Up Platform.

Such a lay-out would not, of course, be favoured to-day, and it is rather extraordinary that it survived as long as it did, for if you take another look at the plan, you will see that no through Down train could pass the station if an Up stopping train was arriving and, in fact, only one train could have access to the station at a time.

With its obvious disadvantages from an operating point of view, however, this lay-out of the through and platform lines was followed at other large stations such as Taunton

Sonning Cutting to-day

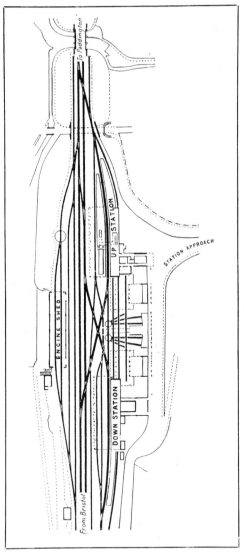

Diagram of Reading Old Station

Reading Station about 1895

and Exeter, and appears to have worked satisfactorily for the requirements of the period.

Incidentally, Reading was not only the first, but also the last, of these one-sided stations and, although some alterations were made and complications introduced when a branch connection for the Newbury and Basingstoke lines was put in later, the old station survived until the present one was built in 1896.

West End of Box Tunnel

TALK NUMBER FIVE

BOX TUNNEL

Still following the advance of the railway, River Thames was crossed by brick and stone bridges at Basildon and Moulsford and carried largely by embankments and some deep chalk cuttings on to Hay Lane (Wootton Bassett Road) some three miles beyond Swindon, which was selected as the place for " the principal locomotive station and repairing shops " of the whole line. When the railway was opened to Hay Lane in December, 1840, however, no station had been provided for the small market town of Swindon.

Meanwhile, work had been going on at the Bristol end of the line and the section Bristol to Bath had been opened in August, 1840. We will follow the rails westward, however, and come to the Hay Lane to Chippenham section where, in order to preserve the level, four deep cuttings and three high embankments had to be provided. These clay embankments caused a lot of trouble and anxiety by slipping and, in one case, rows of piles had to be driven down through the embankment to the ground below and chains used to hold together opposite rows of piles—an expensive procedure.

And now we approach the most formidable engineering problem on the whole of the original Great Western Railway from London to Bristol—the Box Tunnel. The

Box Tunnel, from an old print

section Chippenham to Box was indeed a difficult one, for a stone viaduct west of Chippenham station is followed by two miles of Lowden embankment and then by nearly three miles of Corsham cutting from 40 to 70 feet in depth through solid stone to the mouth of Box Tunnel. Situated ninety-nine miles from London, the tunnel is nearly two miles in length, and at the time of construction was by far the longest railway tunnel in the world.

Box Hill, through which the tunnel was driven, is at the north-western end of the Cotswolds. It rises to 400 feet above the railway level and is composed of strata of oolite (Bath stone) and fuller's earth in liassic and chalk formation. Quantities of excellent building stone are quarried from this locality and, as you may probably know, it is the same as Portland stone which was selected by Sir Christopher Wren for building St. Paul's Cathedral.

BOX TUNNEL

Four hundred feet of stone right in the way of the railway! Here, indeed, was a solid sort of obstruction, and at first a tunnel was considered no more practicable than a cutting. The proposition was formidable enough to deter engineers less bold than Brunel, but he decided to go through it and *straight* through it, which is the way he treated most obstacles which he encountered—and not such a bad way either, all things considered.

The project came in for a lot of criticism when the Bill for the construction of the railway was in Committee, but although there were numerous objections to the tunnel, they proved on examination to be largely of a sentimental nature; for example, one statement made was that " no person would desire to be shut out from the daylight with a consciousness that he had a superincumbent weight of earth sufficient to crush him in case of accident." It was

Box Tunnel East end

also described as " monstrous and extraordinary," " most dangerous and impracticable tunnel at Box."

An engineer critic, who certainly doesn't appear to have been enamoured of the idea, said he believed " the inevitable, if not the necessary, consequence of constructing such a tunnel would be occasionally the wholesale destruction of human life, and that no care, no foresight, no means " (that he was acquainted with) " that had been ever applied up to that time could prevent it."   Rather terrifying, had his belief been well-founded !

The principal witness against the proposal, however, was a Dr. Lardner, who seems to have been regarded as an engineering authority of some standing at that time. He brought before the House of Lords' Committee elaborate calculations which were intended to prove that, if the brakes were to fail, a train in the tunnel would have acquired a speed of 120 miles per hour.   The Committee were aghast at such evidence from such a savant.   It transpired, however, that Dr. Lardner had entirely omitted from his calculations the restraining forces of friction and air resistance and, as Brunel was able to demonstrate, the speed could not exceed 56 miles an hour !

To illustrate his point, Brunel prepared a model of the cutting at the mouth of Box Tunnel, with the rails running up an incline of 1 in 90.   It had been asserted that not only would the noise of two trains passing each other in the tunnel be so appalling that no human being could stand it, but that the gradient was so steep it could not possibly be ascended, and that in coming down the trains would inevitably get out of hand and " take charge." " Now, my Lords," said Brunel, " kindly point out whether this dreadful gradient into the tunnel is an ascending or a descending one." Owing to an optical illusion they one and all guessed that the rails dipped *down* into the cutting. Brunel had also prepared a little model of a tip-wagon. He loaded it with pieces of chalk and placed it on the rails at the entrance to the tunnel. It rolled gently down to the mouth of the cutting and there tipped, scattering the pieces of chalk over the committee table.

Brunel had a moſt trying examination in Committee, as we have already seen, but I think he muſt have enjoyed himself on that day at any rate.

It is recorded that the Duke of Wellington, who was one of the members of the Committee, was highly delighted with the model Brunel had used and kept asking for more and more chalk wherewith to load the little wagon, which then repeated its firſt performance. Here we see the " Iron Duke " in a new and very human aspeƈt.

Work can be said to have commenced on the tunnel in 1836, when trial shafts were sunk in order to find out the nature of the ground, and contraƈts were later let for six permanent and two temporary shafts—all 28 feet in diameter. In February, 1837, the direƈtors reported that the completion of the permanent shafts, which varied from 70 ft. to 300 ft. in depth, was well advanced.

Contraƈts for the aƈtual tunnelling were then let, and provided for monthly progress of the work and for completion in two and a half years—by Auguſt, 1840. The greater part of the tunnel was to be brick-lined and the remainder left bare—i.e., the more solid rock face to form the roof and sides of the tunnel.

Considerable trouble was experienced from water, and one of the shafts was flooded on two occasions, while steam pumps had to be employed at other shafts to get rid of water, which at one time rose to a height of 56 feet and stopped work at that point.

Work on the tunnel went on day and night, and it is said that the navvies imported for the job took all the available lodging accommodation in the locality. The workmen played (very appropriately) "*Box* and Cox," those on the night shift turning into the beds when the day shift men turned out of them. As many as 4,000 men and 300 horses were continually employed, whilst a ton of gunpowder and a ton of candles were used weekly.

The excavation work was done entirely by manual and horse power ; the horses being employed in walking round and round and turning drums, and thus working the winding gear which hoisted the excavated material up the shafts.

The navvies employed were a roughish lot apparently, for they seem to have given a good deal of trouble in the villages where they lodged and, as this was before the days of police, the 26 inspectors appointed to the tunnel work had also to do duty by turns on Sundays in trying—not too successfully—to keep the peace among these rough elements. It is not surprising, perhaps, to learn that there was much genuine rejoicing among the inhabitants when the tunnel was finished, and the navvies took their departure. Probably, from some of the local residents' point of view, the latter was appreciated quite as much as the former.

Middle Hill Tunnel, Box

Although the work was much interfered with by water getting into the workings, practically two-thirds of the job was completed by February, 1840, and by that time three additional shafts had been sunk and work carried on from them in both directions.

Thirty million bricks were used on the tunnel work, the majority of which were made by a local brickmaker, who had a hundred horses and carts employed near by for three years. The total amount of material excavated was 247,000 cubic yards.

As the work proceeded from each end a good deal of anxiety was caused as to whether the excavations should prove to be in a straight line, but when the workmen broke through the laſt intervening ſection of rock, it was found that the line of the headings was accurate " to a hair." The two roofs formed an unvarying line, whilſt the side walls were within an inch and a quarter, much to the relief and joy of all engaged on the job. This was indeed a highly creditable achievement, and was probably quite without precedent.

The tunnel, which is said to have coſt £100 a yard to conſtruct, even in those days of cheap labour, is 3,212 yards in length, is 30 feet wide at the spring of the arch and the crown is 25 feet above the rails. It is ventilated by six shafts 25 feet in diameter varying in depth from 70 to 300 feet. The tunnel is perfectly ſtraight and ſtanding at one end daylight can be seen at the other. As long ago as April 12th, 1859, there appeared in *The Daily Telegraph* under the heading " Curious Fact," the following :—

" It is a remarkable fact that annually on the morning of April 9th the sun's rays penetrate through the great Box Tunnel of the Great Weſtern Railway, and on no other day of the year " ;

and it is ſtill more remarkable that April 9th is the birthday of Brunel, who was responsible for the conſtruction of the tunnel.

The opening took place on June 30th, 1841, which was a day of rejoicing in the neighbourhood. The tunnel mouth was bedecked with flags, and a decorated train which left Paddington at 8.0 a.m., arrived at Briſtol in four hours, and at Bridgwater (the firſt ſection of the Briſtol and Exeter Line having been opened in June of that year) in five and a-half hours.

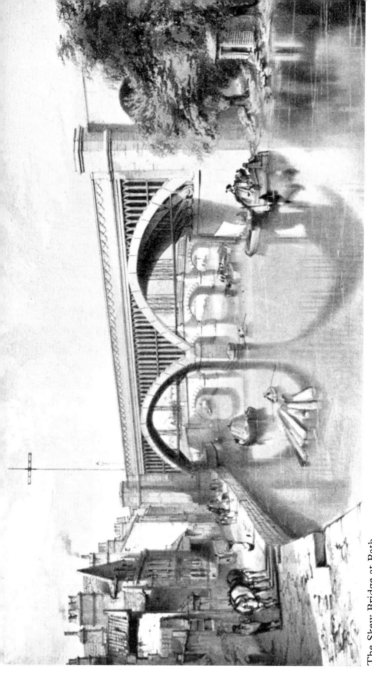

The Skew Bridge at Bath

Twerton Tunnel, East end

Trains were not lighted in those days, and the idea of travelling in darkness through such a tunnel seems to have struck terror into the hearts of some members of the public, if it had not already been implanted there by the terrible and nerve-wracking forecasts made by many detractors of the scheme. So successful had been the alarmists that for some time after the tunnel was opened for traffic, many passengers (observing the spirit of a safety slogan of a century later) would leave the train at one end of the tunnel, drive along the roadway overhead to the other end, and there await another train in which to continue their journeys. In fact, *Keene's Bath Journal* for Monday, July 26th, 1841, contained a public notice heavily underlined regarding the *Star* Coach, which travelled to Reading through Calne and Chippenham daily, to the effect that :—

" Persons fearful of Box Tunnel may go to Chippenham by this coach, and proceed on the line of the railway by the 11 o'clock train."

BOX TUNNEL

An attempt was made to light at least part of the tunnel by reflector lamps, but proved unsuccessful and Brunel made one of his typically " Brunellian " reports on the subject to the directors, which concluded :—

" I am afraid there are no means of remedying the evil of darkness in tunnels (the extent of evil, however, is this, that the tunnel is during 24 hours as dark as the rest of the line frequently is during the night, but is otherwise exposed to fewer casualties), unless by a general and brilliant illumination, which would, of course, be very costly."

You will already have appreciated that Brunel was a remarkable engineer and one who, when he made a decision on any project, went through with it regardless of any obstacles. In Box Tunnel we have an excellent example of that unwavering determination of this Napoleon among engineers.

ᔑ   ᔑ   ᔑ   ᔑ

The section of line from Box to Bristol was a difficult one from a constructional point of view, and embodied some heavy engineering work, although no single problem calls for our detailed examination.

Between Box and Bath there was a short tunnel through Middle Hill (200 yards), the deep Ashley Cutting, and single-arched stone bridges over the Avon at Bathford and just east of Bath Station, which was built on arches with the booking office below it.

Proceeding westwards, the line again crosses the Avon, just outside Bath Station, by a skew bridge, and we must, I think, spare a few words for that bridge which existed until 1878, when it was replaced by an iron girder structure.

Bristol Old Station

BOX TUNNEL

Although the width of the river at the point of crossing is only 80 feet, the skew bridge was of two spans of that width with a central pier of masonry—164 feet in all. Each arch consisted of six ribs (five feet apart) springing from the central pier, each rib being made up of five layers of Memel timber. You can see what a handsome bridge it was by this photograph. The tow path was carried under the western arch by an iron galley.

The skew bridge over the Avon was followed by a viaduct of no fewer than 73 arches over two roads. Then came Twerton Viaduct of 28 arches, a walled cutting, and Twerton Tunnel (264 yards). After another cutting there is Saltford embankment, more than two miles in length over low-lying ground, followed by another and deeper cutting, then by Saltford Tunnel (176 yards) and two other short tunnels.

There are then three more tunnels known as Nos. 3, 2 and 1 (reading westwards) of 1,017, 154 and 326* yards respectively, and before coming to Bristol the line at one point has to be carried on a shelf of rock alongside the river, where a huge retaining wall had to be built.

The approach to Bristol Station is by bridges over the Avon, the Canal, and the Floating Harbour, and the Station itself had, like that at Bath, to be built upon brick-work arches ; but at Bristol both the booking office and waiting rooms were originally below the rail level.

This is, I fear, far from being a full and detailed Story, but you will, I think, appreciate from what I have told you that the Chippenham to Bristol section had more than its full share of problems for the railway engineer.

---

*No. 1 Tunnel has since been opened out (1890).

Cornish Riviera Express passing the Sea Wall at Dawlish

TALK NUMBER SIX

SOUTH DEVON RAILWAY—AN EXPERIMENT IN AIR PROPULSION.

IT is doubtful if the most optimistic G.W.R. enthusiast realised the extraordinary capacity for development and expansion in that undertaking when, on June 30th, 1841, the line from London to Bristol was opened for traffic. Probably its enterprising engineer in his highest flights of prophetic vision, never saw his original Great Western Railway of 118 miles as the undertaking of about 3,800 miles of " first track," roughly lying within the triangle London-Birkenhead-Penzance, which is the Great Western Railway of our time.

I am sure you will be surprised to learn that in the constitution of the Great Western Railway as we know it to-day, no fewer than 255 separate undertakings have lost their original identities. Prior to the operation of the Railways Act of 1921 (when the railways of the country were divided into four groups), the Great Western Railway was the longest railway in the Kingdom—with 2,785 miles of first track. Between the years 1838 and 1890 about 1,690 miles were added to the system, rather by the absorption of small undertakings than by the construction of new lines.

It is somewhat strange, perhaps, to think that only sixty years ago the Great Western Railway proper terminated in the West of England at Bristol. But it did, for the line from Bristol to Penzance was then in the hands of four separate companies.

As we have seen, by 1841 the Bristol and Exeter Railway had been opened (to Bridgwater) and until 1849 this undertaking was worked by the Great Western Railway as an extension of its own, the line through to Exeter being brought into use on May 1st, 1844. Perhaps I ought to make it clear that Brunel was the engineer of this line as he was of the South Devon Railway (Exeter to Plymouth), the Cornwall Railway (Plymouth to Truro and Falmouth), and the West Cornwall Railway (Truro to Penzance), all of which were destined to become part of the Great Western Railway.

The last link in the route, the West Cornwall Line (from Truro to Penzance), was the conversion of a narrow gauge mineral railway from Redruth to Hayle, opened in 1841 and converted to mixed gauge by laying a third rail, to enable broad gauge vehicles to use it, and by extensions Redruth to Truro and Hayle to Penzance.

But we are slipping into history again—let's get back to our engineering features. There was none, I think, which calls for special mention on the 75 miles of the Bristol to Exeter Railway. Brunel kept his line practically level as far as Taunton, and to this day the line from London to Taunton *via* Bristol is the most level stretch of main line track in the country.

Beyond Taunton there is a rise up the Wellington Bank to Whiteball Tunnel (1,092 yards). It was down this bank that an Ocean Mails train from Plymouth to London attained a speed of 102.3 miles an hour in 1904, which was the highest speed achieved by a steam train and a world's record for over thirty years.

I really think we must digress for a moment or two to say a little more on that achievement. It was recorded by Mr. Rous-Marten, a well-known authority and writer on locomotive performances, who took systematic instrumental records of the speeds obtained on the journey.

Incidentally, I would like you to note that the high speed recorded was attained during the run of a train going about its business of carrying mails, and not on a test run specially arranged for the exclusive purpose of achieving high speeds.

It is not, however, the wonderful speed with which we are concerned at the moment, but the fact that, at a speed of over 100 miles an hour the running of the train was (according to Mr. Rous-Marten) " *so curiously smooth that, but for the sound, it was difficult to believe that we were moving at all.*" In an article in the *Engineer* of May 20th, 1904, describing the train run and the record speed attained, Mr. Rous-Marten wrote : " *Here the speed was very high, and, as I have had occasion to note in other cases, the running was the smoothest and the steadiest of the whole journey. The movement resembled a slipping along the smoothest ice, and not the slightest oscillation could be detected in the van where I made my observations.*"

You will, I am sure, agree that, while saying much for Great Western Railway locomotives, the historic achievement of May, 1904 was also a fine tribute to the high standard of Great Western Railway permanent way. It was, of course, a combination of locomotive and track efficiency which made the record speed possible.

73

Collapse of the Sea Wall

By the way, that record was not divulged (except for articles in the technical press) until eighteen years afterwards. The explanation is to be found in the fact that in the early days of the twentieth century the world moved much more slowly than it does to-day, and anything exceptional in the way of a train speed was regarded as " tempting the fates." As a matter of fact, letters frequently appeared in the press in those days protesting at " terrific " or " dangerous " speeds, when it transpired that an express train had travelled on some parts of its journey—as Great Western trains often did, of course—at 70 or 80 miles an hour.

The line from Exeter to Newton Abbot is comparatively level and includes that delightfully picturesque stretch of track—so beloved by holiday-makers to the West Country—along the sea coast from Starcross to Teignmouth and then by the bank of River Teign to Newton Abbot. Just beyond Newton Abbot is the formidable Dainton Bank, which rises for two-and-a-half miles with the sharpest gradients used by main line express trains anywhere in the world. The line falls from Dainton to Totnes, and then rises again to Rattery and beyond to Wrangaton, although from Rattery the gradients are less steep. Just beyond Rattery is Marley Tunnel (869 yards). There is then a gradual descent to Hemerdon and then a steep decline to Plympton followed by a short rise to Mutley Tunnel (183 yards) and then a decline to Plymouth.

We are going to talk about gradients and look at some diagrams directly, so I won't give you any figures as to the rise and fall of the line at this stage.

The line from Exeter to Plymouth had its full share of engineering problems. It involved the construction of a long viaduct 200 yards long over the marshes at Starcross,*

---

*Replaced by a tipped embankment 1898-9.

Thickening of Sea Wall to take thrust of new Arch

and extensive blasting of the face of the cliff from Dawlish to the commencement of Teignmouth Beach to provide a terrace for the track. Tunnelling through projecting headlands was necessary at five points—the longest of these tunnels being through Parson's Rock (374 yards)† and at Teignmouth East Cliff‡, but perhaps the greatest feature was the Dawlish and Teignmouth Sea Wall so well known to visitors to those resorts.

The sea wall is provided to protect the railway at this exposed position, and throughout its early history this section of line suffered from the effects of gales and the buffetings of high seas. In latter years, however, thanks largely to continued vigilance of the Engineering Department, little serious trouble has been experienced.

Several gales occurred when the sea wall was under construction and work was delayed owing to the difficulty in landing material from the sea. A portion beside the Exe estuary was destroyed by storm in 1847 and there was further trouble in the winter of 1852-3 when, due to heavy rains, part of the cliff between Dawlish and Teignmouth gave way and brought down a section of the sea wall, interrupting rail traffic for two days.

During the severe weather of the Crimean winter (1855) much of the beach was actually washed away near Teignmouth, and the soil on which the sea wall was built was scoured out for a distance of 50 yards, with another serious interruption in railway traffic. Temporary arrangements were made and during the next two years the sea wall was entirely rebuilt.

---

†Lengthened to 513 yards in 1920.
‡This tunnel was opened out when the track was doubled in 1884.

Teignmouth end of the Sea Wall

Further damage has occurred due to severe weather from time to time, but the steps taken by the engineers in constructing groynes, and by other expedients, have rendered attacks by the sea on this stretch of line less effective.

The sea wall is composed of limestone masonry behind which is a filling of hard rubble. It extends 2,080 yards and is 45 feet wide, 33 feet being occupied by the double railway track, which replaced the single broad gauge track as laid by Brunel. The railway is separated from the public promenade on the sea wall by a parapet which runs beside the railway for its entire length.

This sea wall structure is a most important one and you will realise its responsibility as a defence more fully, perhaps, when I tell you that the Great Western Railway has statutory control over the material forming the beach for a width of 180 feet seawards from the wall.

There were seven of Brunel's imposing viaducts between Totnes and Plympton consisting of stone piers of tapering

Colonnade Viaduct at Dawlish

Pillar of Colonnade Viaduct at Dawlish

twin pillars (8 ft. at base and 6 ft. at top) with timber superstructures, and this illustration of Ivybridge Viaduct, which was 252 yards long and 114 feet high, though by no means the longest,* will give you some idea of these wonderful viaducts, all of which were replaced by masonry structures, but not before 1893. I have some other photographs of Brunel's famous viaducts and the structures which replaced them which we will look at presently.

Brunel had now carried his lines to a part of the country where he could no longer obtain such levels as he had—often at considerable expense—hitherto, and the necessary gradients on the South Devon Railway doubtless influenced him in making on this line, one of the greatest of all experiments in railway propulsion.

Now you must know that at about this time considerable interest was being displayed in engineering circles on the subject of air propulsion on railways, and in some quarters the highest hopes were entertained of the possibilities of "harnessing the atmosphere" as a motive power. To be strictly accurate, however, the proposal was not one of air propulsion, but of propulsion by lack (exhaustion) of air in a pipe, thus creating a vacuum in front of a piston.

Brunel was keenly interested in the possibilities of what became known as the atmospheric railway system. It was tried out experimentally at Wormwood Scrubs and seems to have worked quite satisfactorily there, for powers were obtained for its adoption on the London and Croydon Railway extension to Epsom.

---

*Blatchford Viaduct was 309 yards in length.

Ivybridge Viaduct

SOUTH DEVON RAILWAY

The system was installed between Kingstown and Dalkey in Ireland, and was so successful that it was proposed to extend it as far as Bray. A deputation from the directors of the South Devon Railway visited the railway at Dalkey and made a favourable report upon it.

Let me try and give you a brief description of the system. A cast-iron pipe of 15 inches diameter, having a continuous slot 2½ inches wide throughout its length on its upper side, was laid between the rails and inside this pipe travelled a piston connected to a long frame, to which was attached an arm which passed through the longitudinal slot in the pipe and connected with the vehicles to be propelled along the rails. The slot in the pipe was closed by a valve consisting of a continuous leather flap, greased to make an airtight joint at one edge and hinged at the other. An arrangement of wheels, attached to the frame, pressed up the flap in advance of the arm, but behind the piston. The air was pumped out of the end of the pipe in advance of the train by stationary engines fixed at intervals along the railway, and the part of the pipe behind the piston was left open to admit air by the pressure of which upon the piston the trains were propelled.

The pumping engines were put into action to exhaust air from the pipe when a train was due, and put out of action (ceased to pump) when the train had passed. I trust that is fairly clear, but here are some diagrams which will, perhaps, make it more so.

Stephenson was one of the engineers who condemned the system and regarded it as being quite uneconomical on steep gradients. Brunel, however, on this matter held exactly the opposite view, for he considered the atmospheric

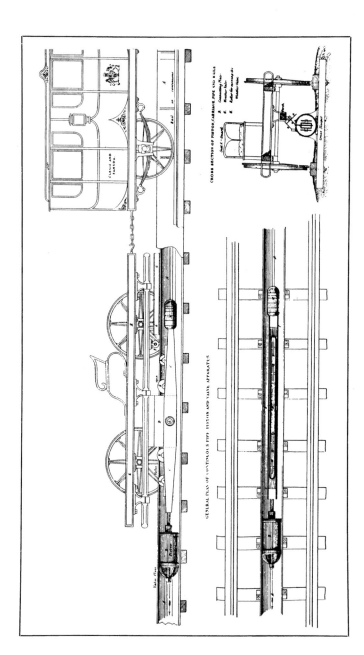

CROSS SECTION OF PISTON, CARRIAGE, PIPE AND RAILS

GENERAL PLAN OF CONTINUOUS PIPE, PISTON AND VALVE APPARATUS

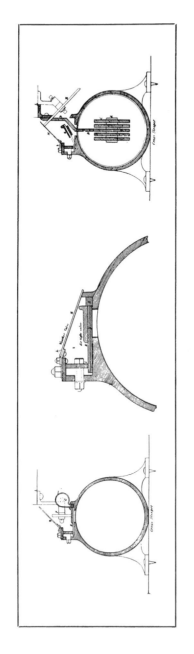

Above—Diagrams showing Elevation of Section of the Atmospheric System

Below—Cross Section details showing Cross Sleeper, Closed Valve and Open Valve system particularly suitable for use throughout the South Devon Railway where its adoption should, in his opinion, enable a single line to meet the requirements, whereas a double track would be needed if the line were worked by locomotives. Brunel estimated that air propulsion would result in savings both in capital outlay and in running costs, as the haulage of engine and tender up the steep grades encountered would be eliminated. He was able to get his scheme sanctioned by the directors, and it was installed and brought into use from Exeter to Teignmouth (between which points eight pumping stations were provided) in February, 1847, and extended to Newton Abbot in January of the following year.

Section of Pipe forming part of original plant installed in 1846 for working the South Devon Railway

Atmospheric Pump-house

TRACK TOPICS

The system seems to have worked quite well—to begin with, at any rate. It is on record that speeds of 68 miles an hour were attained with light trains and 35 miles an hour with a 100-ton train, while out of 884 trains, 790 kept (or gained) time. Passengers liked the smooth running and freedom from all smoke and coke dust.

Unfortunately, however, this highly satisfactory state of things at the start did not last. Defects arose, and the leather valve responsible for keeping the system airtight seems to have been the weak spot, as the leather deteriorated and resulted in air leakage. Despite the renewal of the leather valve over the entire length of the pipe line, the trouble recurred and on occasions brought the trains to a standstill.

To cut a very long story short, the system was maintained on the line Exeter to Newton Abbot for eight months, when Brunel, after a full investigation, was compelled to admit that the system was unsatisfactory and recommended its being scrapped and locomotive haulage substituted.

You may ask whether, seeing that an otherwise satisfactory system of propulsion failed owing to the difficulty in keeping the valve airtight so many years ago, anything has been done in the interim to remedy the defect. That is a perfectly reasonable question, particularly in view of the vast strides which have been made in engineering since those days. Frankly, however, I cannot tell you, but it must be remembered that since that time the steam locomotive has been vastly improved, and electric traction introduced, so that probably it has not been considered worth while to experiment further with air propulsion on railways.

Old Atmospheric Pipes now put to another use

The failure of the system on the South Devon Railway must have been a great blow to Brunel. It was a costly business for the railway and Brunel was a considerable loser financially by the experiment, having himself invested a good deal of money in the apparatus.

The result of the failure of the atmospheric system was that the South Devon Railway found itself with 40 miles of single line track which, but for the atmospheric installation, would have been double, with steeper gradients than had been intended for locomotive working, and a series of useless engine houses and pumping plants distributed along the railway.

Royal Albert Bridge, Saltash

TALK NUMBER SEVEN

ROYAL ALBERT BRIDGE, SALTASH

HAVING taken the railway to Plymouth, the next section westward, Plymouth to Truro and Falmouth (Cornwall Railway), necessitated crossing the estuary of River Tamar, from Devonshire to Cornwall.

The original scheme for bridging the estuary, which is 1,100 feet wide, was by means of a structure with a main span of 255 feet and six others of 105 feet but, owing to Admiralty requirements for a headway of 100 feet, that plan had to be amended.

Brunel then had an ambitious proposal for a bridge with a main span of 850 feet, but finally settled on one with two main spans of 455 feet, supported by a central pier. Robert Stephenson's Britannia bridge across the Menai Straits had two spans of 460 feet supported by a central pier placed on a convenient rock in the middle of the waterway. In the middle of the Tamar Estuary there was no such convenient foundation for a pier, but 70 feet of water over a thick bed of mud—a vastly different proposition, as you can well imagine.

This did not deter Brunel, however, and he proceeded to explore the bed of the estuary in order to provide his own foundation by means of a large wrought-iron cylinder 85 feet long and 6 feet in diameter which, by a framework between two hulks with lifting and lowering tackle, was sunk through the 16 feet of mud and clay to the solid rock below. It was thus possible, by boring, to ascertain the nature of the rock, and the cylinder was lifted and lowered until the borings had been taken over the whole site on which the central pier would rest. This work was carried out in 1849 and, having ascertained what he wanted to know about the rock on the site where the pier would be placed, Brunel set about the complete plans for his bridge.

Constructional work was commenced in 1853 and the most important and most difficult task was building the central pier, sometimes at a depth of 66 feet beneath low water level—no mean undertaking in those early days of engineering. For this purpose, Brunel had designed a cylinder somewhat like that which he had used for exploring the bed of the estuary. This one was 37 feet in diameter and 85 feet long and so constructed (by means of a transverse diaphragm 20 feet from the bottom) as to form a diving bell. The space below the diaphragm was divided into a central circular chamber and round this there were radial spaces, any of which could be used

Cylinder constructed to explore bed of Tamar Estuary

separately as a diving bell. Between the diving bell portion and the top of the cylinder was a central shaft 10 feet wide with a smaller working shaft and air lock inside it. This huge contraption, consisting of cofferdam and diving bell combined, and weighing over 300 tons, was made on the shore close by the site of the bridge, and so constructed that, after the central pier had been built within it, the cylinder could be divided and removed in two parts.

Well, the cylinder was duly completed and floated out from the shore, drawing 50 feet of water. As Brunel reported, it was " pitched with its lower edge accurately, that is, within three or four inches of the exact spot required," in June, 1854. That in itself seems a pretty good achievement, but it was only the first step in providing the central pier, which was, of course, the key to the whole structure.

Meanwhile, work was going on from the shore in providing the land piers, etc., and the locality must have looked like something between an ironworks and a ship building yard with its pontoons, engines, scaffolding, and the construction of the bridge spans, including the two immense main trusses, which were built on the spot.

It does not require a deal of imagination to visualise Brunel, in those busy days, overlooking the erection of the bridge piers or moving about the hive of activity on shore ; exchanging a cheery word of encouragement or joking with those engaged in preparing the great trusses which were destined to carry the railway across the wide expanse of the estuary. We are told that Brunel infused a light-hearted gaiety into his most serious labours and generated

Saltash Bridge during construction

an infectious enthusiasm among all concerned in his great enterprises. Can't you see him wearing that familiar " stove-pipe " top hat and the immense necktie affected by his generation and, of course, smoking the inevitable cigar, as he watches his wonderful design gradually materialising ?

But we must follow that cylinder which had been pitched on the site on which the central pier was to be placed. February, 1855, saw the cylinder—after some excavation work on the rock—finally settled into its correct position, and the work of building up the masonry of the pier inside it was commenced. Sometimes this work had to be carried on against a pressure of 70 to 80 feet of water.

Progress under these conditions was necessarily slow, but with interruptions due to the influx of water (remedied

Walking along the top of the Bridge

Saltash Bridge from an unusual viewpoint

by providing extra pumps), bad weather, and other causes, ten feet of the pier had been completed by the end of 1855. The whole job of building up the masonry portion of the pier (about 12 feet above high water) was finished towards the end of the following year and the cylinder was divided and removed according to plan. The most difficult and hazardous job had been done and the most essential feature of the bridge provided.

Now, perhaps, we ought to turn our attention to the big main trusses which were being built up on the Devonshire shore while the other essential work for the bridge had been going on from each side, and in the middle of the estuary.

The two main trusses are clearly shown in these photographs. Each consists of an arch-shaped member, being a wrought-iron tube, oval in section, 16 ft. 9 ins. wide and 12 ft. 3 ins. high, the ends of which are connected by a suspension chain of link plates. The arch and chain are connected at eleven points in the truss by verticals, themselves braced together by diagonal bars. From these verticals are suspended the girders carrying the roadway. Each truss is 56 feet high at the centre, 455 feet long, and weighs 1,060 tons.

The next task—and no small one either—was the floating of the trusses into position and then raising each of these thousand-ton spans into place 100 feet above the water. No chances were taken with this job, which was well rehearsed beforehand. The western truss was first completed and was lifted from its berth by two pontoons placed one at either end.

Engine driver's view of Saltash Bridge

Interior of one of the main tubes

Brunel controlled the important operation of floating the western truss into position (which took place on September 1st, 1857) by flag signalling from a platform provided in the centre of the truss, like a captain (as indeed he was) on his bridge. About 500 men were at their various posts, and the pontoons carrying the high truss were gently manœuvred by means of hawsers from a number of barges previously placed in position, and from the shore. Out across the river the great truss was drawn and cleverly swung round into the precise position required with its ends over the bases of the central and western piers, so that when water was admitted to the pontoons, allowing these to float away, the truss settled on the pier bases.

The event seems to have been celebrated as a general holiday and gala day throughout the locality. It was a gloriously fine day and it is said that the crowd which assembled to witness the proceedings numbered upwards of 300,000. For its entertainment every provision seems to have been made by the local inhabitants, whose houses were decorated with flags for the occasion.

A description of the placing of the first main truss from a pamphlet issued locally at the time is of interest. Listen to this :—

" Not a voice was heard, not a direction was spoken ; a few flags waved, a few boards with numbers on them were exhibited, and, as by some mysterious agency, the tube and rail borne on the pontoons travelled to their resting-place, and with such quietude as marked the building of Solomon's temple. With the impressive silence which is the highest evidence of power, it slid, as it were, into position without an accident, without any extraordinary mechanical effort, without a " misfit " to the extent of the eighth of an inch. Equally without haste and without delay, just as the tide reached its limit, at 3 o'clock, the tube was fixed on the piers 30 feet above high water, and the band of the Royal Marines, which was stationed in a vessel near Saltash, struck up " See the conquering hero comes," and then " God Save the Queen," when the assembled multitude broke out in loud and continued cheers in expression of their admiration and delight."

What a great day that must have been for Brunel !

It was now possible to raise the truss by means of hydraulic jacks about three feet at a time and build the piers upwards as the truss was gradually lifted. This was, of course, comparatively slow work as the masonry of the land piers had to be given time to set between each lift. The central pier above the masonry is of four octagonal cast-iron columns, built up in sections and connected by cast-iron work. The structure above the railway on the central pier is a cast-iron standard, while that on the shore piers consists of masonry in a cast-iron casing. All the pier superstructures are arched and through these arches the trains pass.

By the following May the western truss had been raised to its final position 100 feet above ordinary spring tides. Two months later the eastern truss was successfully floated to its site, but Brunel was not able to be present on this occasion, and his ill-health was already causing anxiety. By December this eastern truss had been raised to its final lofty position. A good deal of other work on the bridge could not be taken in hand until the main trusses were positioned, but the bridge was completed and tested early in 1859.

I do not think we can omit the following further short extract from the report already quoted from, as it is rather a fine example of the heavy journalistic style of those days :—

" Henceforth old Tamar will be spanned by its double ferruginous bow, presenting with Cyclopean triumph a grand highway of commerce across the broad bosom of the waters."

Presumably the modern equivalent of that " ferruginous bow " would be " a rhapsody in steel," but I am afraid we cannot aspire to such æsthetic heights of expression in these matter-of-fact talks, although Brunel's masterpiece is worthy of the very best that can be said or written about it.

Unfortunately for some time previously the shadows had been gathering around Brunel, and when, in May, 1859, the bridge was opened by the Prince Consort, after whom it was named, its engineer was too ill to be present and had, in fact, only four months to live. Brunel was abroad under treatment at the time of the opening of the bridge, and on his return to England he paid a visit to the West of England, and saw his great work in its completed form for the first and last time : reclining on a couch placed on a trolley (for by this time he was failing), as he was drawn slowly across the wonderful structure, the admiration of engineers throughout the world, which his master-mind had conceived and created.

ROYAL ALBERT BRIDGE, SALTASH

Brunel's friends on the Cornwall Railway afterwards placed on the landspan archways in raised letters the inscription " I. K. Brunel, Engineer, 1859."

The bridge is 2,220 feet in length, and besides the two main spans of 455 feet each there are seventeen land spans varying from 70 to 90 feet each which, on the Cornish side, are on a sharp curve. The height of the centre pier is 240 feet from the foundation and the railway is 110 feet above high water level. The whole structure is an embodiment of engineering skill and beauty of line which might well have inspired the phrase " the poetry of engineering."

The structure of the two main trusses was unique, being a combination of an arch and a suspension bridge—the outward thrust of the arch on the abutments being counterbalanced by the inward drag of the chains. These chains, by the way, had been originally made for the Clifton Suspension Bridge.

There is, I believe, a little confusion about Brunel's bridge chains, so I would like to make the matter quite clear. The facts are these : Brunel's suspension bridge at Clifton was commenced in 1836, but owing to difficulties in raising the necessary capital, work was abandoned in 1853, and chains which had been made for the bridge were utilised for the Royal Albert Bridge at Saltash. In 1860, the year after Brunel's death, the work on Clifton Bridge was recommenced by a company composed of leading members of the Institute of Civil Engineers, who wished to complete it as a monument to their late friend, and the opportunity was taken by Sir John Hawkshaw to acquire the chains from the Hungerford Bridge over the

The construction of Saltash Bridge, depicted by
Mr. W. Heath Robinson

ROYAL ALBERT BRIDGE, SALTASH

Thames at Charing Cross—one of Brunel's structures which was then being demolished—for use at Clifton.

Well, that's the story of the construction of the Royal Albert Bridge, Saltash. I am sure you would like to see Mr. Heath Robinson's idea of how the job was done, and here it is. The drawing is one of a series of " gems of absurdity " by that artist, published by the Great Western Railway under the title " Railway Ribaldry."

ᕙ   ᕙ   ᕙ   ᕙ

What countless train-loads of happy holiday-makers proceeding to and from Cornish resorts must have travelled over the Royal Albert Bridge ! Every weekday afternoon the " Down Cornish Riviera Express " passes over it at about a quarter to three o'clock *en route* to Penzance, and the " Up Riviera " crosses the bridge from Cornwall to Devonshire just about a quarter past twelve (noon) on its journey to London.

Saltash Bridge as it is to-day

TALK NUMBER EIGHT

ROYAL ALBERT BRIDGE, SALTASH— RENEWING THE LAND SPANS

I T is eloquent testimony to the genius of Brunel that, despite the steady increase in the weight of locomotives and other rolling stock passing over it, the Royal Albert Bridge, opened in 1859, remained unaltered (except for the addition of 401 new cross girders in 1905 and the reconstruction of the first two spans at the Cornish end, to enable the double line at Saltash Station to be extended on to the bridge) until 1928, when it was decided to replace the main girders of the remaining fifteen land spans.

This provided an interesting engineering problem, for the physical conditions ruled out all methods usually adopted in bridge reconstruction work. Cranes could not stand and slue on the bridge, and further, owing to traffic requirements, occupation of the single line could only be given for limited periods.

Although it means jumping from 1859 to 1928 at a bound, I think, while we are considering this famous bridge, you might like to know how this problem of renewing the land spans, 70 years after construction, was solved.

The probability of the land spans requiring renewal sooner or later had, of course, been in the minds of G.W.R. engineers for some time past, and a method of reconstruction

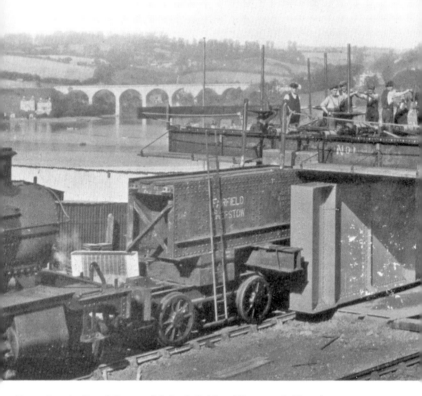

Renewing the Land Spans of Saltash Bridge (Photograph No. 1)

had been approved in principle by the Chief Engineer. The scheme eventually adopted was the use of a specially designed vehicle, which came to be known as the "erection-wagon," of such a size and shape that it would :—

(1) Carry two new main girders, one on each side of it, from the temporary constructional yard near by, to the particular span for which they were made.

(2) Deposit them on their piers, in position between itself and the old girders which they were to replace.

(3) Carry the whole of the floor while, with specially designed gear, it picked up the old girders and deposited them in temporary positions outside the permanent positions of the new girders.

(4) Place new main girders in permanent position.

(5) Pick up the old main girders and carry them to the temporary yard for cutting up.

This vehicle was of lattice design, 95 feet long (in order to bridge the longest span) and weighed about $18\frac{1}{2}$ tons.

Photograph No. 2

Attached to the top boom were transverse girders forming four cross members, and within each pair of cross girders were fixed lifting and traversing gear, all tested for a working load of 15 tons.

Fixed on the underside at each end were box girders, 12 ft. in length (adjustable as their positions would vary on each span) which rested on stools and carried the total weight during the time when the replacement of each of the girders of each span was being effected.

I could give you a long and detailed description of this ingenious erection-wagon, but what is far more helpful for our purpose, I think, is this series of four photographs taken at various stages of the work.

I might add, however, that the erection wagon was provided with two standard engine bogies, with special wooden bolsters. The total weight of the wagon when carrying the old and new girders, etc., was approximately 80 tons.

There was, of course, a certain amount of preliminary work before bringing the erection-wagon on the scene at all. This consisted chiefly of the providing of close-decked suspended scaffolding under each of the spans, laying a longitudinally supported track in the place of the former cross-sleeper road, separating (by means of oxy-acetylene plant) the original cross-girders, and so forth.

The actual job of replacing the land spans was carried out on Sundays when the traffic was comparatively light. The times of certain through trains were temporarily re-arranged, and the local services to Saltash were run between Plymouth and St. Budeaux stations whence passengers were conveyed by road motors to the ferry across the Tamar. Absolute occupation of the main line was obtained for limited times, during which the original cross

Photograph No. 3

girders were removed, the old timber decking renewed, and the new main girders unloaded from the crocodile wagons on which they arrived from the makers, and placed in position alongside the erection-wagon in the siding. On the Sundays when the main girders of each span were renewed, the occupation of the line extended from about 9 a.m. to 2 p.m.

The general procedure followed in renewing a span was this : the erection-wagon, with the new main girders suspended therefrom, was drawn slowly out of the siding by a locomotive on to the main line as shown in the photograph marked " No. 1 " and thence propelled on to the bridge over the span to be renewed. The steel stools of the erection-wagon were then fixed in position over the piers and the erection-wagon girder itself was lowered on to them.

The bogies of the erection-wagon having been run clear of the work, the permanent way and cross girders were then attached to it and the new main girders were lowered on to temporary packing on each pier, thereby removing their weight from the erection girder. The erection-wagon was then lifted bodily by four 40-ton hydraulic jacks and set on cross timber beams inserted on the top of the stools, this lift raising the cross girders sufficiently to clear their supports on the new girders and to draw the shackles tight on the bottom flanges of the old main girders.

Everything now being clear, the four men detailed to each of the four main screws (with one additional man to check the horizontal and vertical movements to ensure all working together), lifted the old girders and moved

Photograph No. 4

them outwards to positions where they would clear every-thing when drawn off the bridge. The new main girders were then travelled outwards (look closely at photograph No. 2) and lowered to permanent position; the bridge floor, which had been suspended from the wagon, dropped on to them, and the permanent way restored.

Photograph No. 3 shows this stage of the operations, the old main girders are suspended from the overhead traverse girders and (as you can see) the " wagon " end is being transferred to its bogie preparatory for the journey back to the construction yard, and photograph No. 4 was taken during this journey.

This procedure describes the method of renewing all but the two innermost land span girders, one end of each

Saltash Bridge from Cornish side

of which was housed in the casing of one of the main piers, and to enable these to be withdrawn and replaced by new girders, a travelling platform was designed and fixed at each end of the erection wagon ; after the old girders had been lifted as previously mentioned, they were transferred to the traverser, drawn out of the portal and connected to the main screws, and the new girders run into the final position by similar means.

During the periods between the Sunday occupations, the work of cleaning, painting and riveting proceeded. The whole of the reconstruction was carried out by the Great Western Railway Company's staff, the average number of men employed on each Sunday being about 46. The average time taken in reconstructing a span was about three and a half hours.

ROYAL ALBERT BRIDGE, SALTASH

Now we must go back about 70 years again.

While the Royal Albert Bridge at Saltash had been under construction, progress was being made with the Cornwall Railway, and by the time the bridge was completed in 1859 the route to the West was open from London to Truro.

The most notable engineering feature of the railway from Plymouth to Truro and on the Falmouth Branch (Truro to Falmouth), apart from the famous bridge, was the large number of viaducts, carrying the lines over the many valleys with which this broken country is seamed. Between Plymouth and Truro, the aggregate length of the 34 viaducts is no less than four miles.

In making railways through hilly country it is generally found that cuttings form the material for embankments, but where there are few cuttings there is, of course, a lack of such material, and for economy reasons, as in Cornwall and to some extent in South Devon, viaducts were built to bridge the valleys.

Brunel's viaducts were almost a unique feature in railway construction and, apart from those crossing low water or marshy ground (where all-timber structures were erected), his viaducts consisted of masonry piers 66 feet apart carried up to 35 feet below rail level. From each pier four sets of heavy timbers radiated upwards fanwise, and carried three longitudinal beams making a flooring six inches deep. Diagonal braces and tie rods were added to minimise deformation.

Timber was employed largely for economical reasons, which also influenced Brunel in adopting a more or less standard design. It is said that these lofty spider-like structures inspired west-countrymen with something akin

Truro Viaduct as originally built

After reconstruction

to awe. This is not altogether surprising, for to people with no previous experience of railways, night travel in anything like a gale at such altitudes muﬅt have been rather an ordeal—at firﬅt at any rate.

The higheﬅt of the viaduﬆs was St. Pinnock (151 feet), with a length of 211 yards, whilﬅt that at Liskeard, only one foot lower, was 240 yards long.

The original conﬅruﬆion has been replaced in some cases by an entirely new masonry viaduﬆ built alongside ; in other cases, where it was not convenient to divert the track, the superﬅtruﬆural work has given place to ﬅteel girder-work.

The laﬅt of the old viaduﬆs to disappear was that at College Wood on the Falmouth Branch, replaced laﬅt year.

The laﬅt railway on the route to the Weﬅt was the Weﬅt Cornwall Railway from Truro to Penzance which, as you have already been told, was partly conﬅruﬆed from a narrow gauge mineral line. This seﬆion presented no outﬅtanding engineering features and the viaduﬆs, which were somewhat lighter in conﬅruﬆion than those on the Cornwall Railway, were all rebuilt between 1871 and 1899 and that at Penzance was in 1921 replaced by a ﬅtone embankment and sea wall.

∽　　∽　　∽　　∽

We have now followed Brunel's railways from London to Penzance and looked briefly, at any rate, at some of the greateﬅt problems which confronted the indomitable engineer on that route. The lines to the Weﬅt of England, of course, are a part only of Brunel's wide field of opera- tions on the Great Weﬅtern Railway and, superficial as has been our survey, we cannot even attempt to cover the vaﬅt

Old Carnon Viaduct shewing piers for new viaduct alongside

G.W.R. system in anything like the same order or detail.

And now, if you charge me with having so far made Brunel the hero of our talks, I must certainly plead "guilty," but maintain in defence that he undoubtedly *was* the hero. It has been said before, and said quite truthfully, that the name of its first engineer could no more be omitted from any history of the Great Western Railway than could mention of King Charles' head be kept out of that memorial of the late Mr. Dick—and these talks of ours must be, to some extent, historical.

Having seen how some of the many engineering problems on that historic route from London to Penzance were overcome by the genius of Brunel, I think it is time we turned our minds to the railway track of to-day, so we will next have a talk on modern railway track components and construction.

Carnon
Viaduct
during
reconstruction
—
New arch
viaduct on
left ; old
viaduct on
right

Reconstruction
completed

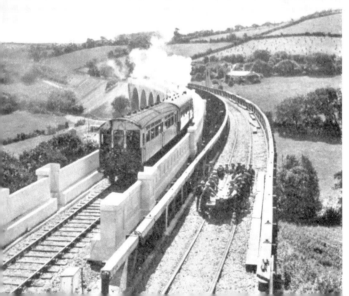

Stacking Sleepers at Hayes

TALK NUMBER NINE

MODERN TRACK CONSTRUCTION

THE faſt, smooth and comfortable " running " of Great Weſtern Railway trains, for which you can vouch, and for which the G.W.R. has so long been famous, is due to four main faƈtors—locomotive efficiency, the conſtruƈtion of the carriage ſtock, the condition of the track, and signalling inſtallation. I think it is true, however, to say that the greateſt faƈtor in providing smooth running is the perfeƈting of the alignment and top surface of the rail.

The broad gauge remained the ſtandard form of permanent way for many years, although the weight of the bridge rail was progressively increased from 43 to 71 lb. per yard. Following a period when 80 lb. flange rails and 80 lb. double-head rails were used on cross sleepers, the firſt heavy bull-head rail weighing 86 lb. per yard was introduced in 1882. From that date bull-head rails in caſt-iron chairs became the accepted form of track, and to cope with the ever-increasing axle loads, the weight of the rail was gradually increased until in 1900 the Great Weſtern Railway adopted rails weighing $97\frac{1}{2}$ lb. per yard. This remained the ſtandard for main lines until 1921, when the British ſtandard rail, weighing $94\frac{1}{2}$ lb. per yard, was introduced.

Only the beſt is good enough for the G.W.R., and this is particularly true of such an important part of its equip-

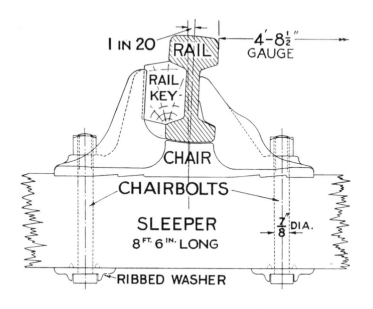

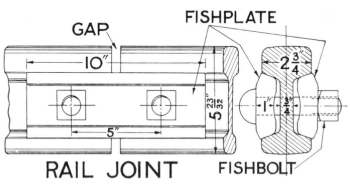

RAIL JOINT

Some details of the Permanent Way

ment as permanent way material. Stringent inspections and tests are carried out to ensure that all materials and components comply with the exacting specifications laid down by the Chief Engineer, and chemical analyses are made in the majority of cases.

You learned something about the components of the permanent way on the occasion of your trip to Swindon,* and now you may like to have information on the subject in a little more detail.

Here is a diagram giving details of the various components of standard G.W.R. permanent way.

The ballast forms a sort of buffer between the track and the " formation," and you can well understand that if this formation or bed was of hard rock without a cushion of ballast, the running would be harsh and there would be a lot of wear and tear on the track components. Another purpose of the ballast, which is composed of either broken slag, granite or other hard stone, or gravel, is to enable surface water to drain away quickly to below sleeper level and so into the drains alongside the permanent way. The lower ballast laid directly on the formation is much larger than the top layer which has to provide an even bed for the sleepers.

You already know a good deal about steel sleepers which are still in the experimental stage, but you may like to hear something more about the wooden sleepers generally in use. These are chiefly of Baltic Redwood and British Columbian Fir, both soft woods, and a much smaller quantity are of Australian Jarrah, a hardwood.

All soft wood sleepers are creosoted before use and each

*See " Cheltenham Flyer."

G.W.R. Sleeper Depôt at Hayes

Baltic sleeper absorbs, on the average, about three gallons of creosote, British Columbian Fir absorbing about half this amount. The Jarrah sleepers, being of hard wood, cannot be effectively creosoted, and are used clean. The average life of a creosoted Baltic sleeper is from 20 to 25 years, and that of a Jarrah sleeper about 20 to 30 years.

The Great Western Railway sleeper depôt and creosoting yard is situated alongside the railway just before you reach Hayes, travelling westwards from Paddington, and there the sleepers are adzed, bored, creosoted, and chaired ready for use. Here are some photographs of these various processes.

All sleepers are machine-adzed to accommodate the chair (which is serrated at the base), and to enable a standard length of chair bolt to be used. The operations of adzing the serrations on the sleepers to match those on the chairs and boring the bolt holes, are performed on the same machine, as you see in the illustration. The adzing drums operating at each end of the sleeper have 64 knives in all, and the drums make 3,000 revolutions a minute. As you can see, the sleepers are fed to the machine and carried by means of a dogged chain under the adzing knives on the one side to the other side, where four holes are bored simultaneously in each sleeper, after which they pass on to the elevator and are from there loaded on trollies for creosoting.

The sleeper-loaded trollies are run into huge cylinders, about 90 feet long and of 6 feet 9 ins. internal diameter, at the ends of which are quick-locking doors. As many as 600 sleepers can be dealt with in one cylinder and when the heavy doors, each weighing about three tons, have

Adzing
Machine

Boring
Machine

Conveyor
from
Machines
to
Creosoting
Trolley

Creosoting Cylinders

Hammering in Chair bolts

Chairing Machine

been closed—which operation takes a man about one minute—air is exhausted from the cylinder until a vacuum of about 23 ins. is registered. Creosote, which is an oily tar liquid, heated to a temperature of about 150° Fahr. is then drawn up into the cylinder from the tank below in which it is stored, and by means of a pressure pump the creosote is forced into the sleepers until the requisite quantity has been injected, the pressure, which varies for different classes of timber, can be brought up to 200 lbs. per square inch. After the pressure has been released, any creosote not absorbed by the sleepers is run back into the tank below. A vacuum is then created in the cylinder for about fifteen minutes to take away superfluous creosote from the timber, and such creosote is also dropped in the tank below. Trollies containing the creosoted sleepers are then run out on to drainage pits which collect any further surplus creosote which may drip off. The whole process of creosoting takes from an hour to an hour and a quarter for Baltic sleepers.

For chairing, the creosoted sleepers are placed six or eight at a time on a bolting table where the chair bolts are hammered home. The sleepers are then tipped over, one at a time, on to an electrically-driven dogged chain conveyor which transports them to the chairing machine. On the way there the chairs are dropped on the projecting bolt ends and the nuts threaded on and given a couple of turns.

Each sleeper is pushed by the dogged chain conveyor into the chairing machine (take a look at the photograph), and the operator opens a valve which causes two hydraulic rams to descend and engage the jaws of the chair, setting them to gauge and pressing the chairs on the sleeper with

124

a force of ten tons on each ram. While thus held, two operators, by means of a lever, pull down four revolving spanners, which quickly tighten up the nuts. The spanners of this machine are driven by friction wheels which allow them to slip when the nuts have been screwed up sufficiently. This ingenious labour-saving machine can chair 120 sleepers in an hour.

After leaving the chairing machine, each sleeper is carried along electrically driven conveyors until it is dropped on special " crocodile " sleeper wagons, each of which holds 160 chaired sleepers. The load is then secured ready to be sent to whatever part of the line it may be required.

Perhaps I ought to explain that the object of the bolt thread being on top is to minimise corrosion which, as you can readily understand, would be increased if buried in the ballast. This also has the advantage that the amount of tightening can be more readily judged.

What is that ? Oh yes, I ought to have explained that the serrated base of the chair is to prevent lateral movement or " spread " of the chair on the sleeper and thus take the strain off the bolts and assist in maintaining the correct gauge. You may be interested to learn that as a result of this provision it is found that, even in the case of sleepers which have been in fast running lines for upwards of twenty years, no " ovalling " of the holes in the sleepers could be detected, which proves that practically no variation in the gauge of the track had taken place.

The standard G.W.R. main line rail (which is the British standard) weighs $94\frac{1}{2}$ lb. per yard and the weight as supplied by the makers is not allowed to vary more than one half per cent. above or below the normal. The standard lengths of rails are 45 and 60 feet, the latter now being more generally used.

Rails are rolled from steel made by the Bessemer Acid, Open Hearth Basic or Acid processes, and you may be surprised to know that it requires $2\frac{1}{2}$ tons of iron ore, $3\frac{1}{2}$ tons of coal, and 8 cwts. of limestone (6 tons 8 cwts. of material in all) to make one ton of steel for rails.

A chemical analysis is made of every 100 tons of rails rolled to see that they conform to the specified limits, and one of the physical tests consists of dropping a ton weight on a 5-feet length of rail (placed on bearers giving a span of 3 ft. 6 ins.), first from a height of 7 feet followed by another drop from 20 feet. The " set " is measured after each blow and must not exceed the prescribed limits.

Each rail is also examined for straightness, freedom from burrs, or other visible defects, and the length has to be

Standard
18-inch
Fishplate

New
10-inch
Fishplate

within $\frac{3}{16}$ in. of the normal, whilst the fish-bolt holes (for the fish-plate bolts which join the ends of lengths of rail) must, of course, be accurate to gauge and centres.

The rails are joined together with fishplates 18 ins. long, each single joint having one pair of fishplates and four fishbolts. The latest improved rail joint is the short two-hole fishplate 10 ins. long needing only two fishbolts with a correspondingly reduced spacing of sleepers on either side of the joint, as has already been explained to you.*

Clearance is provided between the fishbolt and the fish-bolt hole in the rail and (though you probably would not think so) the fishbolt holes in the rails are not round like the bolts, but oval, so as to permit the movement of the rails in expansion or contraction due to changes of temperature.

*See " Cheltenham Flyer."

127

TRACK TOPICS

In joining up the ends of rails, the gap has to be varied according to the season of the year—from $\frac{1}{4}$ in. in summer to $\frac{1}{2}$ in. in winter—for standard length rails. It is those rail ends which are the cause of that familiar beat or rhythm made by the train in passing along the track.

I am afraid you would find details of all the various tests of the track components somewhat wearying, but I may add that fishplates, chairs, bolts, etc., are all subject to exhaustive tests in order to ensure that no faulty material passes into service.

This brings us to the last component in a standard section of track—the wooden key which is inserted between the rail and the outer jaw of the chair. Teak keys are used in nearly all main passenger lines as they give a much longer life than oak keys ; the latter are, however, used in less important branch lines and sidings. The keys used to be of two kinds, right-hand and left-hand, but are now reversible. You will see by the diagram how they are driven in the direction of the traffic.

Whilst a tapered key is more efficient than a parallel key, since there is a reserve for tightening, the amount of taper must be strictly limited, for if too pronounced, the key is liable to shoot back under passing loads. The standard wooden key, which is 6 ins. long, has only $\frac{1}{16}$ in. taper, and that proves highly satisfactory. The requirement of a good key is maximum contact with both rail and chair to ensure maximum grip and to minimise " creep."

Perhaps you did not know that trains as they pass along the track tend to drive the rails in the direction in which the trains are travelling. This movement of the rails is known as " creep."

" Creep " occurs more particularly on falling gradients and on sharp curves and, curiously enough, the right and left hand rails do not " creep " to an equal extent. From time to time, therefore, the rails have to be pulled back, and various expedients have been adopted in order to check this forward movement. One is by anchoring back the rails at various points by " anti-creep " appliances, and another is by the use of steel " keys."

The principal disadvantage of wooden keys is that in dry weather they are apt to shrink and become loose and a defective key may result in " creep." Steel keys are not liable to shrinkage and, owing to the form in which they are manufactured, they exert a more constant pressure between the rails and the jaws of the chairs. The results obtained to date indicate that the steel key forms a good anti-creep device. The cost of steel keys, being higher than that of wooden keys, precludes their general adoption, but the saving effected in the rectification of " creep " more than balances the increased cost.

In order to counteract the effects of centrifugal force, and so ensure smooth running, the outer rail on a curve is raised above the level of the inside rail, the sleepers being packed higher at one end than at the other in order to give the necessary cant. The amount of cant—or super-elevation as it is sometimes called—depends upon the radius of the curve and the speed of trains passing over it. Owing to the existence of sharp curves in a few places, trains have to be restricted in speed when passing over such curves, and also through junctions where it is sometimes impracticable to provide the necessary cant.

The maximum super-elevation permitted on the G.W.R. is 6 ins., but in practice it seldom exceeds $5\frac{3}{4}$ ins. Super-elevation is generally attained gradually from the straight and level by what is known as a transition curve in which the cant is increased by about one inch in every 66 feet. On high speed lines the cant is increased very slowly, in some cases as little as one inch in 120 feet.

What's that ? . . . (" What exactly is a transition curve ? ") Well, to be perfectly frank, I was rather expecting that question. I must warn you, however, that the term is one difficult to define adequately in a few words, and any definition must sound a bit involved. Still, we'll see what we can do with it. . . . Let's take it slowly and I think you may be able to follow it.

Here we go then (*andante*) : A transition curve is an easement approach between straight uncanted track and a circular curve canted for the appropriate speed, or between two circular curves of " the same hand or flexure " (which is the engineers' way of saying " curving in the same direction "), the radii and cants of which vary fairly considerably. The curvature on the transition is varying throughout its length from infinity at the tangent point with the straight track to the radius of the circular curve at the tangent point where it joins that curve. Throughout the length of transition the cant is being uniformly increased from zero on the straight to the full cant required on the circular curve. The principles are the same when dealing with circular curves curving in the same direction.

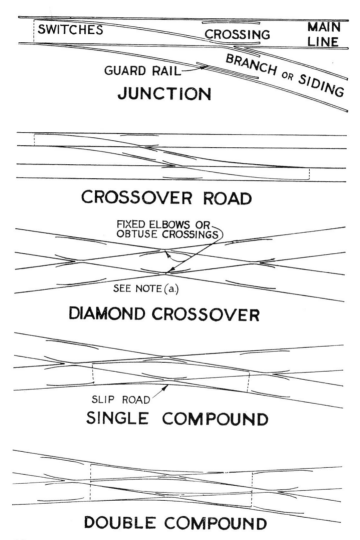

SWITCHES CROSSING MAIN LINE

GUARD RAIL BRANCH OR SIDING

JUNCTION

CROSSOVER ROAD

FIXED ELBOWS OR OBTUSE CROSSINGS

SEE NOTE (a.)

DIAMOND CROSSOVER

SLIP ROAD

SINGLE COMPOUND

DOUBLE COMPOUND

(a.) WHEN ELBOW POINT RAILS ARE MOVABLE THEY ARE OPERATED FROM THE SIGNAL BOX IN THE SAME WAY AS SWITCHES.

Movable Diamond Crossing

Well, that's the definition, and, as I warned you, it is a rather wordy and intricate one.

And now regarding the rails, it only remains to add that they are tilted towards one another by means of the chair seating at an inclination of about 1 in 20. This brings the centre line of the rail at right angles with the tyres of the wheels.

So far we have been concerned with the main components of an ordinary section of main line track, known as " plain line," but you will readily understand that when we come to points and crossings, long timbers have to be employed in place of the standard sleepers.

I do not think we shall do better than take a look at a few diagrams of crossing work, for these are practically self-explanatory.

As you already know, points, or switches, are the means by which a train travels from one pair of rails to another,

132

and consist of movable pairs of hinged raiis, tapering at the points, inside fixed rails. Crossings are the means by which rails cross one another, and their construction has to permit of the wheel flanges passing smoothly over the rails crossed. I should like you to study these diagrams for a few minutes, when I think the purpose of the construction of the crossings will be quite clear. You will see that guard rails are provided opposite the crossings to ensure that the wheel flange takes the proper side of the crossing nose. . . .

Take a look at the diagram of a diamond crossing, and see if you can follow its purpose. In some cases movable elbow crossings are provided, and we have an example of this at Old Oak Common West Junction where the Birmingham line leaves the main G.W.R. line. The elbow points in such cases can be moved from the signal box just like switches, so that they present an unbroken track over the crossing.

Spring Crossing

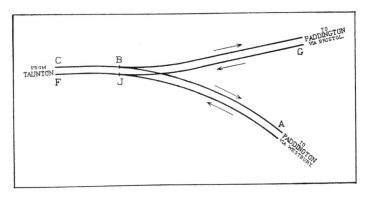

Diagrams of the Junction at Cogload shewing, above, as it was before the construction of the " Fly-over " bridge ; and below, the new lay-out shewing the " Fly-over " bridge carrying the Bristol to Taunton line over the Taunton-Westbury main lines.

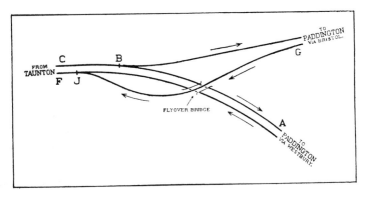

There are several places where the traffic is confined almost entirely to one road through a crossing and in such cases the spring crossing can be used with advantage and economy. It forms a continuous support to the wheels

134

Cogload Fly-over Junction

on the principal road, and is forced open by wheel flanges of vehicles passing along the other (less used) road, the spring returning the crossing to the normal position.

If you think the matter over you will realise that the design and number of double track junctions on a section of railway must considerably affect the number of trains which can be worked over it. Most railway junctions are of the " flat " type, that is, the branch line joins the main line on the level. Now this means, of course, that the " Up " branch line has to cross the " Down " main to reach the " Up " main, or the " Down " branch line has to cross the " Up " main to join the " Down " main according to the direction in which the junction is laid. This, as you can well understand, may cause considerable delays and the remedy is the provision of a " flying " junction.

There are two forms of the flying junction, the " over " and " under." With the former, the " Up " or " Down " branch line, as the case may be, joins its corresponding main line by means of a bridge over both main lines, or the main line it would otherwise cross on the level ; while with the latter, the same result is achieved by burrowing under the main route.

Of the two types of " flying " junction, the " fly-under " (better known as a " burrowing " junction) is generally (but not always) the better, as the speed acquired by running down the slope is counteracted by climbing the succeeding gradient, but with the " fly-over " the incline to be negotiated is, of course, followed by a falling gradient, of which it is not always possible to take advantage.

The provision of a " flying " junction is a very expensive expedient on account of its construction under the amount of additional land required. An excellent example on the Great Western Railway of the " fly-over " type has been provided in recent years at Cogload Junction, near Taunton, where the Down line from Bristol is carried over the main lines to and from London, thus permitting four trains to run simultaneously through the site of the former " flat " junction, which operation is possible on account of four running lines existing on the Taunton side of this junction.

ᔕ   ᔕ   ᔕ   ᔕ

Well, this has been a long talk, and now, I think, you ought to have quite a fair knowledge of the construction of the permanent way. In our next talk we will see how it is kept up to concert pitch.

TALK NUMBER TEN

MAINTENANCE OF TRACK

THE Great Western Railway Engineering Department comprises eleven railway divisions, each of which is controlled by a divisional engineer, who is in charge of a geographical mileage ranging from about 250 to 450 miles, with a length of track including sidings of roughly from 500 to 1,000 miles.

Each railway division is sub-divided into districts in charge of permanent way inspectors, the usual number being ten per division. The extent of a permanent way inspector's district varies according to the nature and importance of the lines maintained, and convenience of supervision, the general average being 38 geographical miles and about 90 miles of single running track, including sidings.

Permanent way inspectors' districts are further sub-divided into about fourteen lengths each maintained by a length gang, comprising ganger, sub-ganger and a number of lengthmen, governed by requirements. The more usual extent of a length is two and a half geographical miles or about six miles of track including sidings.

Briefly, the permanent way inspector is responsible for the examination and maintenance of the lines and all structures and works in his district, the supervision of ordinary maintenance, and renewals and new works allotted to him.

Ganger's Motor Inspection Trolley

Special examinations of bridges, viaducts, and tunnels, and the supervision of repairs and reconstruction works which, in the opinion of the divisional engineer are beyond the scope of the permanent way inspector, are allotted to a bridge inspector, one of whom is employed in each division with a staff of bridge repairers, etc.

Similarly, special examinations of stations and buildings beyond the scope of the permanent way inspector is assigned to an inspector of mechanics. This inspector is in charge of the divisional repair shops, and has under his supervision craftsmen and labourers ranging from 80 to 300, the number being dependent mainly on the works in progress. That is, very briefly, the organisation of the Engineering Department.

MAINTENANCE OF TRACK

Each ganger or, in his absence, the sub-ganger, is required to patrol and inspect his length daily so that every part of the running lines, over which passenger trains pass, and their structures come under examination every twenty-four hours, and any defect which arises is promptly detected and steps taken to make any adjustments or repairs that are necessary.

Particular attention is given to the " top " or surface, as also to the alignment of the rails, whilst the drainage and ballasting call for equal attention. The guiding pressure of engine flanges cuts away the head or top table of the rail on curves ; the sudden approach of hot weather may cause the rails to expand, resulting in track distortion. Such points as these require to be noted and reported, so that prompt action may be taken.

Preparations have to be made against slips or landslides and any tendency in that direction noted, while in certain low-lying districts sudden floodings may and do frequently

Ganger's hand-driven Inspection Trolley

occur and expose the tracks to risk of considerable damage, so that during periods of abnormal rainfall special measures have to be taken for dealing with such emergencies.

Blizzards accompanied by heavy snow-drifts bring their worries for the railway engineer, as you can well appreciate if you take a look at these photographs.

Where any timber comes in contact with sea water, the engineers have to be on the outlook for attacks from a destructive little fellow who goes by the name of " toredo navalis." This little creature eats into the timber, boring a little tunnel as he goes. It is a curious fact that the tunnels never connect, but either go over or under one another. Here is a photograph of a " toredo," who was " caught in the act," and a section of timber attacked from which you can see how destructive this little pest can be when he gets busy.

Chairbolts and keys must be systematically tightened and kept in effective condition. A loose chair causes a springy sleeper and the latter produces a slack, which again quickly develops a crook in the alignment, and if neglected would soon result in unsmooth running.

The method of " packing " the sleepers, with which you will probably be familiar, has been employed since the earliest days of railways and is known as " beater " packing. It consists of driving the ballast under the sleepers with a beater made like a pickaxe, one end of which is flat and the other pointed. Experiments have been made with pneumatic tamping or packing machines in which a petrol-driven compressor operates a number of tampers, but although the work is done speedily by

Abnormal weather conditions must be dealt with

The " toredo navalis" and its ravages

this means, the results are no better and the cost no less than beater packing by hand.

There is another method known as " fly-packing " in which ballast is not driven under the sleeper, but the track is lifted up and a suitable quantity of fine ballast is sprinkled under the sleeper bed by means of a special flat

Beater
Packing

Fly Packing

Track
liners

Slewing
the
track

shovel, and the track then lowered on to this. By this means an even bed is formed upon which the sleeper rests and becomes quite consolidated after the passage of traffic.   About 1,250 miles of G.W.R. lines are now packed in this way, and the results obtained to date appear satisfactory.

MAINTENANCE OF TRACK

Slewing of the track to restore correct alignment is usually done by a number of men using straight bars as levers. Some of the gangs are too few in number to undertake this method, and they are provided with mechanical appliances called " track liners," one of which is placed under each rail and the operation is carried through by two men—as you see in this photograph.

Maintenance methods, particularly on single branch lines, have been much improved by the motor trolley system, and about 25 per cent. of the route mileage of the Great Western Railway is so maintained. Under this system several length gangs on branch lines are grouped into one large gang and supplied with a petrol-driven rail trolley to enable the men and materials to be moved quickly over their length, and the ganger is provided with a smaller car for use on his inspections.

Such branch lines are provided at about every mile with occupation key instruments by which the ganger can obtain occupation of a portion of the line for the trolley or for work at one of these points or, of course, at a signal box. The line is subsequently freed for traffic when the trolley has been removed therefrom, or the work finished, and the occupation key inserted in the key instrument. This is similar to the electric staff arrangement. The motor trolley system enables gangs to maintain much greater lengths of track and thereby effects considerable economies.

I am afraid I cannot even enumerate the hundred and one day-to-day jobs in regard to the maintenance of the

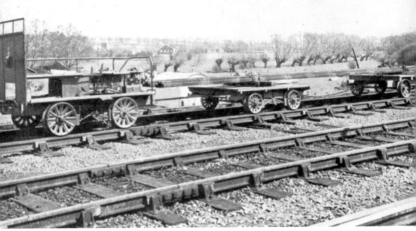

Petrol-driven Trolley and Trailers

railway track, but you will, I feel sure, be interested in the repair of the permanent way by electric welding.

The wear which takes place on the top of the rails, due to the passage of traffic, is much more severe at crossings where there is a break in the continuity of the rail. This results in greater abrasion of the vee-point and wing-rails (and these terms are self-explanatory if you look at this photograph) than of the adjacent rails. Where the traffic on one road of a crossing is heavier than on the other, the crossing naturally becomes worn on one side whilst the other side is still in good condition.

To remedy these defects it used to be the practice to replace the worn parts, viz., the vee-point and wing-rails, with either new or good second-hand material. This entailed not only the cost of the material and manufacture of the vee-point and wing-rails and the labour of taking out and replacing these in the track, but also a good deal of interruption of the traffic; as temporary occupation of the line was required for the work, and crossings are busy traffic points.

Portable electric welding and grinding equipment

Grinding by carborundum wheel

Building up a worn vee-point

In a number of cases the modern method is to build up the worn vee-point and wing-rails on the spot by means of electric arc welding. The plant consists of a light-weight generator, driven by a four-cylinder petrol engine, and engine and generator, when assembled, can, if there is no room at the side of the line, be placed in the " six-foot " way, i.e., between two running lines, without fouling the load gauge. The total weight is 1,200 lb. and the machine can be quickly divided into three parts, each capable of being handled by four men, thus enabling it to be loaded on to an ordinary rail trolley for transport to site, where it can be unloaded and re-assembled in a few minutes.

A portable grinder is worked off the plant and consists of an electric motor driving, by means of a flexible shaft, a carborundum grinding wheel of 8 ins. diameter.

The new metal is added to the worn crossing by passing an electric current through an electrode, or metal rod, held almost in contact with the crossing. The current forms an arc from the end of the electrode to the crossing, fusing the end of the former, and causing the metal of which it is made to flow and adhere to the crossing. When enough metal has been added to the worn parts, any excess or rough edges are smoothed down by the portable grinding machine, held in position by the operator, as you see in the photograph.

The electric arc emits rays well above and below the visible spectrum, and as these rays are harmful to the skin and eyes, the welder is provided with protective equipment. Heavy gauntlets of fire-proofed horsehide protect his hands and arms, while his face and head are shielded

Worn wing-rails before—

—and after electric welding

by a hand screen having a dark glass window which prevents the passage of harmful rays.

In addition to the daily examination of the track by the ganger of each length, and frequent inspection by various officers of the Engineering Department, mechanical means are adopted for ascertaining the condition of the track. Tests are carried out from time to time over the main lines to detect any track faults, some of which may not be discernible on the ground, even to the practised eye.

The track is tested by means of a specially equipped coach used as a track indicator. The " riding " of this coach under standard conditions has been proved by previous tests to be consistently good, and can therefore

Hallade Recorder in use

be taken as a definite indication of the state of the track. The behaviour of the coach is recorded by a Hallade Recorder, a portable instrument by which faults in the track are indicated on graphs, made simultaneously by pendulum-operated pens resting on carbon and tracing papers, the latter papers slowly revolving on a drum operated by a clockwork motor. The coach is also equipped with a speedometer.

When the riding of the coach is disturbed in a transverse direction more than a certain amount, a valve, operated electrically, of an electrical apparatus is opened and a splash of whitewash is deposited upon the track. This splash marks approximately the position of the disturbance and duly receives the attention of the engineering department staff.

When whitewash is dropped a hooter sounds in the coach, and at the same time a mark is made on the Hallade record or graph by an electrically operated pen fitted to the recorder for this purpose. The " whitewash van," as it is known, works to a scheduled programme over the principal lines of the G.W.R. system, attached to the rear of express trains, and the records obtained provide valuable data for the engineers responsible for the maintenance of the track on the various routes.

The use of the apparatus is just one of the many efforts consistently being made to ensure the maintenance of that comfortably smooth running which those who " Go Great Western " take as a matter of course.

So far we have only considered the running lines, but there is also a vast mileage of track composed of lines in

A hump marshalling yard, looking from the hump to the sidings

railway yards and depôts, docks, private sidings, carriage sidings, marshalling yards, etc.

We have said nothing about reception sidings, sorting sidings, or marshalling yards in previous talks, and perhaps a word or two here may not be out of place. Reception sidings are provided to accommodate trains of goods and mineral wagons clear of the running lines whilst waiting to be dealt with in a marshalling yard. The number of such sidings depends upon the volume of traffic received over and above the capacity of the sorting sidings.

The term " sorting sidings " is practically self-explanatory, and these form the main part of a marshalling yard, the number depending roughly upon the number of destinations to which trains are made up and dispatched.

Marshalling yards may be defined as gathering grounds or collecting areas, where wagons for a variety of destinations are received, sorted, made up into full train loads, and forwarded on a further stage of their journeys, or direct to destinations. These yards are of two kinds; flat yards and gravitation yards, and in the case of the former, practically the whole of the work is performed by shunting engines without any assistance from gravitation.

Gravitation yards are roughly of two kinds : (1) Where the natural lie of the ground enables the sidings to be situated on a falling gradient and the force of gravity used to sort the wagons ; the level of the line falling from the reception sidings to the sorting sidings, and (2) where an artificial " hump " is provided over which the wagons are propelled and allowed to run, by gravitation, to their respective sidings. Entry to the sidings is controlled in both kinds of yards by points operated by shunters or from a control cabin. In the case of " hump " yards care has to be exercised in regulating the speed of the wagons over the " hump " in order to avoid one wagon, or two or more coupled together, from overtaking others before clearing the points giving entry to the sorting siding.

In some " hump " yards, hydraulic rail brakes or wagon retarders are installed and by this means the speed of wagons can be checked on their way to the sorting sidings.

I am afraid these brief remarks about marshalling yards come more properly under the head of railway operation, but the maintenance of the tracks is, of course, a matter for the Engineering Department.

You are now pretty well versed in track construction and maintenance, and perhaps you would like to know something about track relaying. Very well then, that shall be our next subject.

Loading and unloading rails

TALK NUMBER ELEVEN

RELAYING OF TRACK

I N the middle of each year a programme of relaying and resleepering to be carried out in the succeeding year is drawn up for the approval of the Chief Engineer. This programme is based on recommendations made by the divisional engineers in view of the age and condition of the track under their charge as further determined by examination on the ground, weighing of rails, etc.

Relaying or renewal of track complete is generally effected when the condition of the components, as a whole, is such that they no longer present the necessary margin of strength for comfortable running of the heaviest locomotives and trains that use the line.

Resleepering is undertaken where, based on traffic requirements, the rails are stout enough to outlast a second sleeper life. The term " re-railing " is self-explanatory, and this is only carried out to a relatively small extent.

A system of classification of lines, mainly based on traffic considerations, has been laid down for general guidance where relaying, intermediate resleepering, relaying with second-hand rails, etc., should usually be adopted.

In the case of crossing work, which includes switches and crossings, and the large " crossing " timbers supporting them, proposals for relaying are also submitted to the Chief Engineer and included in the annual programme. Standard fittings are used as far as practicable, and detailed diagrams are prepared for all complicated " fittings " and those other than standard. Practically all crossing work is constructed of new materials.

You will get some idea of the extent of relaying when I tell you that the Great Western Railway programme for the year 1935 embraces 390 miles of track renewals, the principal materials required including :—

29,000 tons of steel rails.	15,500 tons of chairs.
4,000 tons of bolts.	650 tons of fishplates.
4,300 loads of crossing timbers.	600,000 sleepers.
	1,650,000 wooden keys.

185,000 cubic yds. of ballast.

On the Great Western Railway a novel and economic method of loading and unloading rails has recently been adopted. This is done over the end of the rail wagon and is illustrated in these two photographs.

In the unloading operation the principle is that the rails to be unloaded are anchored, two at a time, to the track behind the wagon, and then the engine pulls the wagon away, leaving behind the two rails which drop on to the sleepers.

Loading rails from the track to the wagon is done by putting the brake on the wagon on which the rails are to be loaded and passing a hauling wire from the locomotive over the wagon on to the rails behind the truck. The locomotive then moves forward and pulls the rails (over a roller on the back of the wagon) into place on the wagon.

As a 60-ft. rail weighs over 15 cwt. and 24 men would be required to lift one, you can appreciate that moving rails by hand is a somewhat laborious process.

Rail-lifting appliance

A rail lifting appliance is also in use for lifting rails from the side of the track into position in the chairs. The arrangement consists merely of a lever, working on wheels running on a light steel frame, high enough to enable the rails to be lifted over the chair jaws and long enough to permit of the lever being drawn back on its wheels sufficiently to traverse the rail sideways. Six men can carry out the operation without difficulty, and you see them doing it in this photograph.

Mechanical aids for drilling and cutting rails, and for boring sleepers at site, have been introduced. An interesting example is a machine which consists of a diminutive petrol engine giving an output of nearly 4 h.p. at 3,500 revolutions per minute. It will operate one drill, or one rail-cutting saw, or one auger for boring sleepers. Though this is the smallest self-contained power unit in use by the Engineering Department, it can drill a rail in about two minutes, a job which takes 20 minutes to do by hand, and saw through a rail in about ten minutes, which takes nearly an hour by hand.

8 h.p. Petrol-electric
Boring and Drilling Plant

Boring Sleepers
with 4 h.p.
Portable Plant

Rail-
cutting
Saw

RELAYING OF TRACK

A more imposing machine is the 8 h.p. petrol-electric boring and drilling plant, which produces electric power at site for operating four augers for boring the large timbers under the switches and crossings, or two drills for rail drilling.

Special relaying gangs are employed for carrying out renewals of any considerable lengths of track. The materials required are conveyed to the site beforehand and unloaded alongside the track—rails, chaired sleepers, fishplates, bolts, keys, switches, crossings, crossing timbers, etc.—and every preparation that can be made beforehand is made, as relaying generally requires complete occupation of the line, and that occupation must be as brief as possible. Relaying is generally undertaken on Sundays when traffic is normally much lighter than on weekdays, and frequently as much as a mile of track is renewed at a stretch on a single Sunday.

When carrying out preparatory work before the Sunday on which the old track will be replaced; for example, when the old ballast has been partly removed in readiness for breaking the road on the Sunday to effect renewal, traffic has to be slowed down to fifteen miles an hour and notice boards are exhibited. A green horizontal board like the arm of a distant signal, but pointed at the end opposite the " swallow tail," showing green and white lights side by side at night, is placed about half a mile to the rear so as to warn engine drivers to reduce speed before coming to the renewal site. Two other track signals, a letter " C " on a stand at the point where the restriction commences and a similar " T " for the point where it terminates, are also erected.

Side tipping ballast wagons

When complete occupation is obtained, the work of relaying has to proceed " at the double," for all traffic during absolute occupation of one line of a double track is worked over the remaining line in accordance with the Company's special regulations for single line working. On branch lines of single track, where the traffic is usually fairly light, relaying is generally carried out " between trains " on weekdays, i.e., without interference with ordinary traffic.

The relaying train, with men, tools and other equipment, arrives on the scene on the Sunday, and from the moment after the last passenger, goods, or mineral train has passed over the section of line to be relaid, it is handed over completely to the Engineering Department, and the old road is attacked. Keys are knocked out, fishbolt nuts

Tipping " screw-over " side tipping wagons

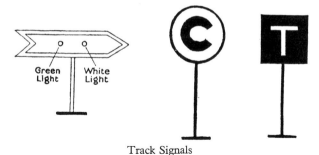

Track Signals

run off, the bolts knocked out and fishplates removed, the rails being lifted out of the chairs and sleepers finally pulled up.

As soon as a sufficient length of old track is removed, and the sleeper beds " picked over," new sleepers (already chaired as we have seen) are placed in position, and the new rails lifted up and tipped into the chairs. Small metal gauges are temporarily inserted between the rail ends to ensure correct spacing for expansion, fishplates and fishbolts are fitted and the bolt nuts screwed up.

The work goes on apace, and the inspector, as he moves up and down the track, consults his watch, for he knows that the road will be wanted for traffic at a certain time, and he is more than a little anxious if things are not going exactly to schedule.

When the last key in the new track has been driven home and the fishplates secured, the track is slewed to a fair alignment and the sleepers packed to obtain a reasonably good rail " top." The inspector then certifies the track as suitable for resumption of traffic at a reduced speed of fifteen miles an hour.

TRACK TOPICS

During the week following the Sunday on which the relaying has been carried out, the track is put into exact alignment and packed, new ballast being dropped from a special ballast train of hopper wagons. A plough is fitted to a van which is run over the length, thus spreading the ballast dropped in the " four foot."

After each sleeper has been thoroughly packed throughout the renewed length, all fishbolts tightened, correct alignment, cant, and " rail-top " obtained, and all clearances to structures such as bridges, platforms, etc., carefully checked, the length is inspected by the divisional engineer who, if he is satisfied with the condition of the track, authorises normal speed to be resumed.

The complete renewal with new materials of a mile of main line track (plain line) costs approximately £2,200.

Making a cutting

TALK NUMBER TWELVE

KEEPING THE TRACK LEVEL (1)
Cuttings — Embankments — Tunnels — Mining Subsidences

As we have already seen, the level of the railway is obtained through high ground by forming cuttings, or driving tunnels, and across valleys by throwing up embankments or building viaducts. Excavation is a very costly item in railway construction, and involves not only digging out material from the site it occupies, but also finding a suitable deposit for it, and the nearer to its place of origin the excavated material can be made good use of, the lower will be the cost. It is often possible in the case of new line construction to utilise the earth removed by making a cutting at one point in forming an embankment at another. The aim is always to make excavation and embankment balance as far as possible.

In the old days all excavation work was, as we have seen, performed by hand labour. Nowadays the mechanical excavator operated by a steam, or an internal combustion engine running on petrol, paraffin, or " heavy oil " used in conjunction with side-tipping wagons, does the job and the cost works out at about half that of manual labour.

An example of the engineering requirements in construction of a modern railway is provided by the short

Cutting under construction—Temporary Tracks for Mechanical Excavator

suburban line between Ealing and Shepherd's Bush, London. This line, which was constructed and is owned by the Great Western Railway, forms an extension of the Central London Railway and provides a through service between Ealing and Liverpool Street. Over practically its entire length of four miles the railway is either in deep cuttings or on embankments, but only one-third of the material excavated to make the cuttings could be used to form embankments. The construction of the line necessitated the removal of the side of the cutting east of Ealing Broadway Station and replacing it by a heavy retaining wall, while no fewer than nineteen bridges had to be built, fifteen of which cross the railway.

The chief trouble with cuttings and embankments is the recurrence of slips. These are not unexpected in new construction until the sides of the cuttings and banks have had time to consolidate. You will not need to be told that when the material excavated is loaded into wagons and taken to the site where an embankment or other filling is

164

required, and there tipped, it takes up much more space than before excavation and takes some considerable time to settle.

The angle of slope of an embankment or cutting varies according to the material of which it is composed. With chalk or rock cuttings the sides may be nearly vertical, but with clay they have to be much flatter and the slopes of banks may be much steeper in tipped rock than in clay or loose earth.

Many of the deepest cuttings are approaches to tunnels, and some of the longest embankments terminate in the abutments to viaducts. We have an example of the latter in the mile and a half of embankment forming the approaches to Wharncliffe Viaduct, and of the former in the deep cuttings leading to Chipping Sodbury Tunnel (South Wales and Bristol direct Line) which are two and a half miles in length at the Bristol end, and a mile in length at the Badminton end. The approach cutting to White-ball Tunnel (between Wellington and Burlescombe) has a maximum depth of no less than 115 feet above the rail.

Embankment under construction

TRACK TOPICS

You already know something of Sonning Cutting two miles long between Twyford and Reading, and about eight and eleven miles farther westward respectively there are good examples of deep chalk cuttings of one mile in length, the sides of which are very steep and about 60 feet in depth at Pangbourne and Goring.

Near Parson's Tunnel at Dawlish the cutting on the up side of the line is almost vertical, rising to a height of over 200 feet above rail level, whilst on the down side the beach level is about 30 feet below the rail.

The construction of the direct London to Birmingham line involved some extensive cutting and embankment works. At Ruislip, Gerrards Cross, and Beaconsfield, there are cuttings with a length aggregating three and a-half miles, and depths up to 85 feet. The cutting approaches to Wheatley Tunnel (between Princes Risborough and Kennington Junction) reaches 90 feet in depth, and the Wheatley embankment, which is 78 feet in height, extends for nearly a mile. At Ardley, in the Ashendon and Aynho section of the line, there is a cutting of no less than five miles in length, having a maximum depth of 45 feet.

We have seen that in his efforts to keep the railway level, Brunel did not spare expense in the construction of either cuttings, embankments, or viaducts, but we must go to the mountainous districts of Wales to see railway constructional problems concentrated in a comparatively small area.

On the old Cambrian Line (now part of the Great Western Railway) we get examples of cuttings, embankments, viaducts, exceptional gradients, and sea defence works ; in fact, practically every engineering problem is represented. The only straight run of track on the whole line is that from Harlech to Talsarnau, and it extends for four and a-half miles only and is practically level.

There is a rock cutting at Talerddig 360 yards long with a depth of 70 to 80 feet, which is 693 feet above sea level, and the line there is on a gradient of one in fifty-two for three and a quarter miles. An embankment at Trawscoed is 1,300 yards long with a height varying from 60 to 160 feet and here there is a gradient of one in forty-two extending for four and a quarter miles.

From Pwllheli to Aberystwyth the line is never far from the sea, and at Harlech the track for approximately half a mile is laid on benching twelve to twenty feet above shore level, and by a continuous masonry and concrete wall, the cliff on the land side rising to a height of 120 feet above rail level.

At Vriog the railway track for one and a-quarter miles is on benching 90 feet above the shore level, and here the main road is carried on similar benching about 100 feet above the level of the railway.

Clay is the cause of considerable trouble to the railway engineer as we have already seen, and clay slips are due to excess water acting as a lubricant. The first step in dealing with this kind of trouble, after taking any immediate remedial measures, is to locate the cause so as to prevent an extension or continued movement of the slip.

The causes may be many, and open ditches at the top

167

of cuttings, broken or defective pipe drains, garden allot-
ments or making of garden ponds by adjoining owners are
among them. Any of the above may cause a slip in the
sides of cuttings, and in some cases the building of a wall
to hold the slip and prevent its extension is advisable.
This remedy, however, calls for considerable expenditure,
as in the construction of such walls it is necessary to take
them down to the undisturbed earth well below the slip,
and they have to be of sufficient strength to stand the
pressure of the displaced earth.

Wherever possible construction of a retaining wall is
avoided, as unless the drainage of the slip is also undertaken, there is a possibility at some later date of the wall
being subjected to greater pressure than it was designed
to stand.

The stabilising of a slip is undertaken by the construction
of a trench through its centre so as to release imprisoned
water and provide a drain by which further water can
escape. These trenches are filled with large lumps of
slag or stone, and are described as " counterforts." Their
purpose is to strengthen the slope against further movement, and by this means the lubricant causing the clay to
slip is removed.

Slips on embankments may be due to waterlogged
embankment formation, as when an embankment is formed
on sidelong ground. In such cases improvement is usually
effected by providing surface drains and drains at the foot
of the embankment on the high side, so as to intercept
the ground water and prevent it percolating in and under
the embankment.

KEEPING THE TRACK LEVEL

When the depth of formation below the surface makes the cost of excavating a deep cutting, with its necessary slopes, drains, etc., approximately equal to the expense of tunnelling, a tunnel is usually preferred.

The depth at which a tunnel is substituted for open cut is usually about 60 feet in firm soil, but the suitable depth varies with the nature of the strata to be traversed, for where rock is reached a deeper cutting, with steep slopes, might prove more economical than a tunnel. In unstable strata, where slips might be apprehended, tunnelling might be resorted to at a much less depth, the level of the line being, of course, arranged accordingly. There are, of course, other considerations, such as the cost of purchase of a wide strip of land and the cutting up of a single property, which may be an important consideration from the landowner's point of view. There is at least one case on the Great Western Railway where the landlord would not agree to a deep cutting dividing his estate, which was in the midst of good shooting cover, and a tunnel had to be provided.

Where the tunnel is through hard rock it is sometimes possible to avoid lining of any kind, but in all other cases it is usual for a substantial brick lining to be provided throughout the tunnel.

It is usual to make an annual inspection of every tunnel for the purpose of determining the state of the brick lining, or where lining does not exist, the state of the rock.

For the examination of tunnels a brake van is used, on top of which is fitted a hand rail along the centre and guard consisting of iron tubing and standard fixed at both sides as a protection. The guard is not a fixture so far

Two
methods
employed
for
examining
Tunnels

as height is concerned and can be moved up and down, depending on the clearance in the tunnel. The men making the examination with the divisional engineer or the assistant divisional engineer, stand on top of the van, each armed with a hammer and what is called a probing iron, which varies in length according to the height of the tunnel. As the van is moved slowly through the tunnel the roof and side walls are probed and tapped to ascertain their condition. Scaling is generally found in patches, and this is probably due to the different qualities of the masonry used in various sections of the tunnel.

All tunnels cannot be inspected from the tunnel van; for instance, in the Box Tunnel, where the unlined portion of the roof is about 40 feet above rail level, a 16-ton crane

Cement grouting in Severn Tunnel

Shoring up a bank affected by mining subsidence

is used, a structure in the shape of a crow's nest being fixed on the top of the jib, capable of holding four or five men (one man with an acetylene flare lamp sits on top of the jib to supply the necessary light) the jib being raised and lowered as required.

A modern feature of tunnel repairs is that of grouting (i.e., filling up joints with liquid cement) existing masonry or brickwork under compressed air, and considerable economies are effected by this process.

Occasionally tunnels have been opened out into cuttings for various reasons, including subsidences, caused by colliery workings, etc. As there will probably be no more fitting opportunity I ought, perhaps, at this point to say a word or two about mining subsidences, for the worries

172

of the railway engineer are considerably increased in mining areas owing to the liability of subsidences, due to extraction of coal and other minerals.

Cases occur where it is necessary to take down arches or heighten abutments of girder bridges in order to obviate a dip and to obtain the necessary headway. In other cases permanent bridges have been removed and temporary structures on trestles erected during the time the subsidence was active. In other instances existing arches have been supported by centering during the continuance of a subsidence.

From the commencement of mining operations until the full effect of the subsidence has been experienced, the railway company's mining engineers inspect the workings and furnish reports from time to time, particularly as to the depth of settlement anticipated, and the likelihood of sudden movement. The reports enable the necessary

Another effect of a mining subsidence

Steel frame-work, Cockett Tunnel

steps to be taken to maintain the line. Where necessary, watchmen are engaged night and day over the length affected and restrictions of speed are enforced.

An interesting illustration of the effect of mining upon a tunnel is provided at Cockett, near Swansea. At the close of the last century serious distortion of the tunnel occurred and a steel framework was placed inside to hold up the brick lining with ribs at three feet intervals, and concrete backing to cope with the dislocation of the arch. This blocked the passage of loads of the standard maximum gauge westwards of its eastern end, and in due course a portion of the tunnel was opened out to remove the restriction.

The subject of mineral workings under railways is an extensive one, and we cannot do more than look at it quite briefly, but I think I have said enough to indicate that it adds in no small measure to railway engineering problems on the Great Western Railway.

KEEPING THE TRACK LEVEL

Here is a list of some of the principal tunnels among the 187 on the Great Western Railway, and you will notice that all included are a mile or over in length.

PRINCIPAL TUNNELS ON G.W.R.

Name of Tunnel.	Length. Yards.	Situate between	Section of Line.
Severn	7,668	Pilning & Severn Tunnel Jcn.	Bristol & South Wales Union.
Chipping Sodbury	4,444	Badminton & Chipping Sodbury.	Bristol & South Wales Direct.
Rhondda.. ..	3,443	Blaenrhondda & Blaengwynfi.	Rhondda & Swansea Bay.
Box	3,212	Corsham & Box	Main Line.
Merthyr	2,497	Abernant & Merthyr.	Merthyr Branch.
Llangyfelach ..	1,953	Felin Fran & Llangyfelach.	Swansea District Lines.
Caerphilly ..	1,941	Cefn On Halt & Caerphilly.	Cardiff & Rhymney.
Wenvoe	1,868	Creigiau & Wenvoe.	Cadoxton & Trehafod.
Sapperton ..	1,860	Coates & Chalford.	South Wales Main Line.
Patchway ..	1,760	Patchway & Pilning.	Bristol & South Wales Union.

Before we leave the subject of railway tunnels you will doubtless like to hear the story of the Severn Tunnel, and this is obviously the place to tell it. It is rather a long story (for construction of the tunnel was a long job), and will be the subject of our next two talks.

Severn Tunnel, Monmouthshire side

TALK NUMBER THIRTEEN

SEVERN TUNNEL (1)

DOUBTLESS the most difficult engineering achievement of the whole of the Great Western Railway system was the construction of the tunnel which passes under the estuary of the River Severn. It is the longest under-water railway tunnel (4 miles 628 yards) in the world, and it attains a depth of 30 feet below the deepest part of the river bed. It has been said that its construction, which occupied about thirteen years, was " outmatched by no experience in the long account of human endeavour."

Prior to the tunnel being brought into use, the most direct route to South Wales for passenger traffic was by crossing the Severn by steam ferry between Passage Pier in Gloucestershire and Portskewett Pier in Monmouthshire, where the river is about two and a quarter miles wide, but the heavy mineral and goods traffic to and from South Wales all had to take the circuitous route via Gloucester.

The scheme for the tunnel was sanctioned by Parliament in 1872 after it had been under discussion for many years. The engineer for the works was Mr. Charles Richardson, who had been a pupil of Brunel and had worked under him on the preliminary construction of Clifton Bridge, and on the railway between Swindon and Cirencester. As resident engineer of the Bristol and South Wales Union

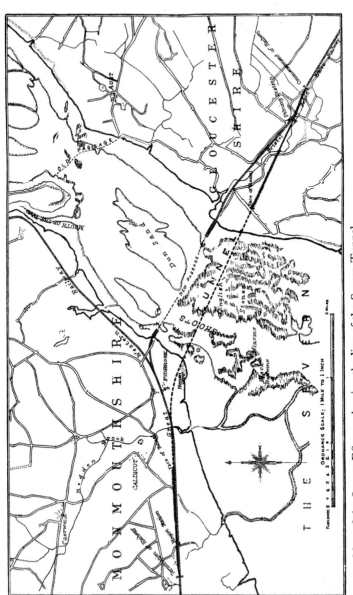

Map of the Severn Estuary shewing the line of the Severn Tunnel

Railway, including the steam ferry across the Severn, he had special knowledge of the currents and the river bed, and it was during the erection of the ferry piers that he arrived at the conclusion that a tunnel under the estuary was a feasible proposition. Mr. Richardson has another claim to be remembered by " boys of all ages " who owe allegiance to " King Willow " for (though it is, perhaps, not generally known) he designed and made the first cricket bat with a spliced cane handle.

It would take many hours to give you an account of the construction of the Severn Tunnel as it deserves to be given, for it is an epic story of conflict with natural forces, but the following is a much abridged account of this great engineering achievement.

The Severn estuary has an extraordinary range of tides, the rise and fall being as much as fifty feet, the highest in Europe, and there are powerful currents. Near the Monmouthshire side of the river bed is a deep water channel, known as the " Shoots," a quarter of a mile wide and having even at low tide a depth of 58 feet of water. As the tunnel was designed to attain a sufficient depth to ensure a roof of 30 feet between the tunnel arch and the lowest point of the river bed, the existence of the " Shoots " meant a deeper and longer tunnel than would otherwise have been necessary ; longer, because of the desirability of avoiding gradients exceeding 1 in 100. In effect, the tunnel had to go as low as 140 feet below the general level of the rails at each side of the river.

In view of the hazardous nature of the undertaking, and the possibility that large fissures might exist in the rock bed of the river, through which the tidal water would

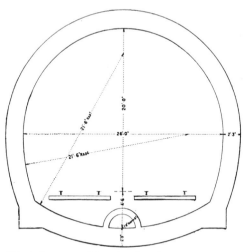

Typical cross-section shewing design of Tunnel

surge in to the tunnel, a clause was inserted in the Act authorising the scheme, to the effect that it could be abandoned if the cost of construction and of excluding water should much exceed the estimates.

Work was commenced in 1873, but owing to landsprings encountered and consequent floodings of the workings, it was not completed until 1886. As many as 3,628 men were employed on the tunnel at the busiest time.

The original intention was to sink shafts, rising to 12 feet above high tide level, in the estuary on either side of the " Shoots " through the rock laid bare at low tide, but covered at high tide by 25 feet of water, flowing in strong currents. This idea, however, had to be abandoned owing to high cost and objections to the interference with navigation which would result. The alternative of sinking shafts in the line of the tunnel, but outside the reach of tidal water on both shores of the estuary, about two miles apart, was adopted.

From each of these shafts, headings seven feet square were to be driven which, as excavation proceeded, would gradually approach one another and then meet, forming a continuous passage and a small preliminary tunnel along the full course, the floors of the heading and ultimately enlarged tunnel being at the same level.

The Seven Feet Heading was to be extended by hollowing out the rock and earth above and alongside to the full cross-sectional dimensions of the tunnel as shown in this diagram. The tunnel, as you see, was designed to take two lines of railway of narrow (now standard) gauge, the width being fixed at 26 feet; the height from the invert was $24\frac{1}{2}$ feet; that is, 20 feet above rails, which were to be $4\frac{1}{2}$ feet above the invert.

After an extensive programme of preliminary work had been carried out, including the taking of soundings over the area of the river bed and fixing on the centre line of the tunnel, etc., work was commenced in March, 1873 (with Mr. Richardson in charge as chief engineer, and Sir John Hawkshaw acting as consulting engineer), by sinking a shaft lined with brickwork 15 feet in diameter.

This shaft, known as the Old Shaft, situated on the precise line of the tunnel, was carried to a depth of 200 feet near the Old Roman Camp at Sudbrook on the Monmouthshire side.

Just here may I be permitted a short diversion by saying that Sudbrook marks a crossing of the Severn estuary used in the days of the Roman occupation ? Only about 100 yards south of the Old Shaft can still be seen the remains of the camp of the Roman warriors who guarded a ferry on the route between Caerleon, which was then the capital of Britannia Secunda and the headquarters of the Augustan Legion, and Bath.

In sinking the shaft a 66-feet strata of hard red marl was encountered at 20 feet and at 45 feet a spring of water yielding 12,000 gallons an hour, necessitating a steam pump. Lower still another spring yielding 27,000 gallons per hour was met with which required the installation of a Cornish beam engine to keep the working clear of water. At about 90 feet Pennant sandstone was struck, and ultimately ironstone.

The bottom of the shaft was reached and the Seven Feet Heading following the axis of the tunnel south-eastward begun in December, 1874, and proceeded until, by August, 1877, 4,800 feet of the heading had been driven under the river bed mainly through Pennant sandstone in which copious springs of fresh water were met. A second shaft was also begun at Sudbrook to form a pumping shaft which by that time was found to be necessary. This shaft was duly completed and (except for the bottom 10 feet) was completely lined with iron.

Progress was such that by October, 1879, five shafts, from four of which seven feet headings were being driven, had been provided—four on the Monmouthshire side and one on the Gloucestershire side. The original Seven Feet Heading from the Old Shaft had been extended for nearly two miles and that from Sea Wall Shaft (Gloucestershire side) to a point where only 130 yards remained between the working faces.

SEVERN TUNNEL (1)

Then a catastrophe occurred.

On the fateful 16th of October, the Great Spring broke into the workings from a subterranean reservoir in such volume that it defied all efforts to dam it back. The workings at the Sudbrook shaft were flooded and the miners engaged, warned only just in time, narrowly escaped with their lives by means of the Iron Shaft. A few hours later the water was 150 feet in the shafts!

That was the first acquaintance with the Great Spring, the persistence of which has daily required a high-powered pumping plant with which the works and later the Tunnel had to be equipped. The quantity of water still raised by the tunnel pumps daily is between sixteen and twenty-five million gallons, and it is estimated that about two-thirds comes from the Great Spring!

Where did all the water come from? You may well ask, for it was *fresh* water which entered the headway in such a flood. The answer was found in the fact that the River Neddern, whose waters join the Severn near by, ran

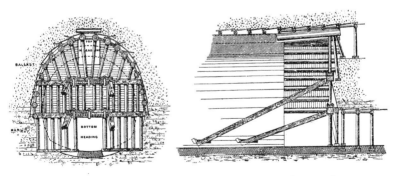

Completed Timbering for full-sized Tunnel in soft ground

183

dry after the irruption for a distance of five miles and many springs round about dried up. North of Sudbrook is a district of about 50 square miles forming the watershed of the Neddern River, which flows through a marshy area under which is a layer of limestone dipping towards the Severn estuary. The waters from the uplands percolate through fissures in the limestone in subterranean streams to the natural reservoir, and this barrier had been broken when the rock at the face of the Seven Feet Heading became too thin to resist the pressure any longer.

The irruption of the Great Spring and all it meant to those responsible for the tunnel work after nearly seven years' labour was a disaster indeed, but their determination to proceed with the undertaking was not shaken. It led to the consulting engineer, Sir John Hawkshaw, being put in charge of the work with Mr. Charles Richardson as joint engineer and in a contract for the completion of the whole undertaking being let to Mr. T. A. Walker, who had been responsible for a part of the work before, and had been associated with Sir John Hawkshaw in constructing the Metropolitan and other railways.

It was not an inspiring job to tackle, but Mr. Walker believed in his ability to see it through. With the workings flooded, pumps idle, and workmen entirely absent, things certainly did not look very rosy. The immediate problems to be faced, however, were means of raising the water from the workings more rapidly than it entered, and damming the source of the disastrous flooding, and these were tackled.

The waters from the Great Spring were eventually shut out by means of two heavy oak shields made to fit the entrance to the Seven Feet Heading on either side of the Old Shaft and placed in position by divers. This was only accomplished after great difficulty due to breakdowns of the pumps, and many plucky attempts by the divers, one of whom, Diver Lambert (of whose exploits you will hear more later), was on one occasion drawn by suction against the mouth of one of the pipes and only freed by the action of three men hauling on a rope.

At last the pumps were able to get the water under control, but only after additional pumps were installed.

Well, the story is a long one of disappointments and difficulties courageously met, backed by the will to win through, and it is only possible now to mention some of the outstanding features, and of these Diver Lambert's exploit certainly calls for inclusion.

An emergency headwall had been built in the workings containing an iron door, the closing of which and the screwing down of two pipes passing through the headwall, would shut off the remainder of the heading and impound the water therein. The 9,000-feet length of heading beyond the headwall was receiving water from a number of copious springs and the headwall door had unfortunately been left open when the Great Spring was tapped and the panic-stricken workmen had fled for safety.

The closing of this headwall was decided on, and it was a risky job for someone. It called for a competent diver with a cool head and plenty of pluck. In the modern idiom Diver Lambert " filled the bill." In the first attempt Lambert had the assistance of two other divers, one to stand at the bottom of the water-filled shaft to pass the air pipe, where it bent into the heading, and another

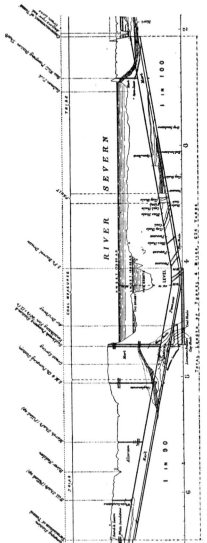

Longitudinal section, shewing geological strata intersected by the Tunnel

to stand half-way along the heading to feed the air pipe forward to Lambert as he moved nearly 1,000 feet encumbered by his heavy diver's equipment.

There he was, alone and a long way from help, proceeding cautiously in inky darkness along a narrow underground passage filled with water, and manœuvring his way past all sorts of obstructions, such as tools and gear and masses of rock, just as the site had been hurriedly left by the workers. Lambert actually got to within 70 feet of the door when, despite every effort, he could drag the air pipe no farther owing to resistance due to its contact with the head timbers exceeding his strength.

There was now nothing for it but to return to the shaft, his object unattained, and the journey back was even more hazardous than that towards the headwall on account of the air pipe curling and getting caught among the timbers overhead, and all these entanglements Lambert had to gather up and carry back with him, for on this, of course, his life depended.

The next attempt was made by a man named Fleuss, the inventor of a self-contained diving equipment consisting of a knapsack oxygen apparatus strapped on the diver's back with pipes to the helmet. Fleuss descended with Lambert to the mouth of the heading, but would go no farther. He failed through inexperience and pardonable nervousness under most trying conditions. When he came up he said the conditions were such that he would not go and close the door for £10,000 !

Diver Lambert then accustomed himself to the knapsack apparatus, made another attempt, and this time reached the door and actually removed one of the pair of tramrails through it, preparatory to closing it when, either by dread of his supply of oxygen becoming exhausted, or by pain due to the long pressure of the unfamiliar equipment on his nose, he again returned to the surface without completing his task.

Two days later Lambert made his third and final attempt ; he again reached the door in the headwall and this time removed the second tramrail, screwed down the valves (as he believed) and returned to the shaft, having been under water eighty minutes. But even then it seemed that the fates were against success for, to the intense disappointment of everyone, those heroic efforts to dam back the water did not result in the lowering anticipated.

By the next day, however, the level had been reduced sufficiently to effect repairs to the pumps and after much trouble with pumping gear it was possible, a month later, for the foreman of the pumps to walk along to the headwall where he found that, owing to one of the valves having a left-hand screw, Lambert had actually *opened* it fully instead of closing it as he had of course intended, and water was still coming through. The foreman closed the valve and it was at once possible to slow down the pumps.

Headwalls of brickwork and cement were then built in order effectively to seal up the Great Spring on both sides and thus shelve the trouble from that source for a time.

And having disposed temporarily of two chief causes of trouble, this would appear to be a convenient point for us to come up from the tunnel for a short " breather."

We will resume and conclude the story in our next talk.

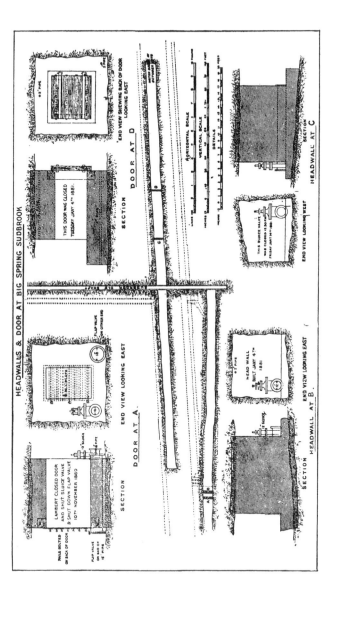

HEADWALLS & DOOR AT BIG SPRING SUDBROOK

END VIEW SHEWING BACK OF DOOR
LOOKING EAST

SECTION
DOOR AT D

THIS DOOR WAS CLOSED
TUESDAY JANY 4TH 1881.

HEADWALL AT C
SECTION

THIS SLUICE VALVE
WAS CLOSED 12.30 PM
FRIDAY JANY 14TH 1881

END VIEW LOOKING WEST

HORIZONTAL SCALE

VERTICAL SCALE

DETAILS

END VIEW LOOKING EAST
DOOR AT A

HEAD WALL
BUILT JANY 4TH
1881

END VIEW LOOKING EAST
HEADWALL AT B

LAMBERT CLOSED DOOR
AND SHUT SLUICE VALVE
& SHUT DOWN FLAP VALVE
10TH NOVEMBER 1880

SECTION
DOOR AT A

SECTION
HEADWALL AT B

Severn Tunnel, Gloucestershire side

TALK NUMBER FOURTEEN

SEVERN TUNNEL (2)

B Y the end of 1880 the flooding of the workings at Sudbrook shaft had been mastered and the Great Spring sealed up so that construction work could proceed.

Additional shafts were now sunk, the pumping equipment augmented, and the work of enlarging the Seven Feet Headway to full tunnel dimensions continued. To supply fresh air to the men working in the headings, air compressing plant was installed. Excavations of rock were effected by drilling and blasting, and compressed air drills were brought into use.

If you look at this longitudinal section of the tunnel you will see that the thickness of roof between the river bed and the tunnel is least at the Salmon Pool on the Gloucestershire side. The intention from the outset was to concrete over the river bottom at this place, and it was here that trouble was unexpectedly experienced by an inrush of sea water at the end of April, 1881. The Salmon Pool is only three feet deep at low water, and to locate the leak men joined hands and waded through the pool. Their search was brought to an abrupt end when one of their number disappeared through the aperture in the bed of the pool, where the water was finding its way into the tunnel. The man was, of course, quickly pulled out none the worse, and the bed of the pool was then overlaid with quantities of clay and specially strengthened brick arching extended past the spot.

A phenomenal snowstorm in 1881 inflicted hardships on all engaged in the workings, one of which was that no coal-supplies were obtainable for the pumping engines, etc., for three days, and the severity of the frost stopped all building above ground for a fortnight.

The next year was a satisfactory one, and good progress was made, but was marred by a considerable fall of coal shale in January with a collapse of the timbering for about 60 yards, but fortunately beds of rock above the shale prevented the water getting into the workings on this occasion.

Although electric lighting was then in its infancy, a 12-lamp system was installed in the Sea Wall shaft in April, 1880, and shortly afterwards a 40-light system on the Monmouthshire side. A telephone line was carried through the heading, giving communication from one side of the river to the other.

Progress was such that by May, 1883, it was necessary to tackle the section through the sealed up area where the dormant waters of the Great Spring were imprisoned. Owing to the great pressure behind the door it was impossible to open it, and after other expedients had been tried, it was decided to drive a bottom heading below the original Seven Feet Heading from the Old Shaft to a point beyond the headwall. A hole driven in the upper heading allowed the water to flow along the lower heading to the shaft and so to the pump.

The Seven Feet Heading was then restored and timbered and the broken door in the headwall made good and another headwall built immediately below it as a precaution, based on apprehensions that further accumulations of water might be tapped.

Unfortunately these misgivings were only too well founded, for on October 10th, 1883, when everything was going favourably, water broke through the face of the lower heading in greater quantities than ever and entirely beyond the power of the pumping equipment to deal with it. It surged through the workings, and a river 16 feet wide cascaded down to the old shaft with a deafening roar.

The doors of the headwall in the Seven Feet Heading and also in the rising heading above it were quickly closed, but it was quite impossible even to approach the headwall in the lower heading. The surge of the water carried the

Sir John Hawkshaw inspecting the Works

men at work in the heading, and even their iron skips and other gear, before it until the stream spread out in reaching the full-sized completed tunnel.

The inrush of water was computed at 27,000 gallons per minute, but lessened until on October 12th the pumps " held " it at 132 feet from the surface. Eventually the pumps gained and the water was down to 13 feet on October 26th. Four days later Diver Lambert was called on yet again, and he went down the lower heading to the door in the headwall and closed it. By November 3rd the pumps had cleared the workings south-east of the Great Spring, which was once more held back by headwalls.

Driver Lambert seems to have had the knack of being on hand whenever exceptionally difficult tasks requiring pluck and fortitude offered. If I may digress for a moment, I should like to give you Mr. Richardson's description of his diver. It is brief, but very pointed. " His name is Lambert, a fair-haired man of few words and great courage."

To resume, on the night of October 17th there was an extraordinary tide in the estuary and a heavy storm. In the darkness a tidal wave five to six feet in height swept over the old sea walls of the Severn (the sea banks for the protection of the tunnel being then incomplete), extinguished the fires of the pumping engines, fell 100 feet down the shaft and imprisoned 83 men in the tunnel below, where the water rose to within eight feet of the roof.

A small boat was obtained and lowered end-on down the shaft and launched on the flood water, and after sawing a way through the timber staging, the men who had been

Tidal Wave—Men in boat sawing through timber staging to rescue workmen

driven back to some higher staging, were at last reached and brought to safety on the morning following the flooding.

The tunnel was actually in a worse plight on October 18th than at any time since the end of 1880 when the Great Spring was first overpowered. The cottages built for the workmen were flooded and huge quantities of timber had floated away, but happily there was no great difficulty in clearing the workings.

By the end of 1883 the tunnel had been completed for 3,774 yards and arching only for 3,179 yards, and the open cuttings of one mile in length at the east and three-quarters of a mile at the west, leading to the tunnel ends, were well under way. The Great Spring was shut off, but the quite incalculable problems it presented were *still* unsolved.

TRACK TOPICS

It was now obvious that yet more pumps were wanted, and these were installed and new engine houses built. This work occupied many months.

At this time the consumption of bricks for lining the tunnel and other works was about a million a month and in order to bring supply up to demand, it was decided to start brick-making on the spot with material excavated from the tunnel. Altogether 76,400,000 bricks and 37,000 tons of cement were used.

Work now proceeded satisfactorily subject to various setbacks which we have no time to mention, the more serious of which were pump failures, but the first nine months of 1884 were generally of fair progress.

Now the time had arrived for the problem of tapping the Great Spring to be faced. It was decided in doing this to divert its waters from the main works by driving a side heading parallel to the original heading, but 40 feet to the north. By September 20th the side heading had been driven 18 yards beyond the site of the headwall, and across the tunnel adjacent to the Great Spring.

Sufficient pumping power being available the sluices in the headwall were gradually opened, when the water flowed down to the pumps. In three days the bottom was sufficiently clear to render it possible to reach the place where the water broke in and fortunately it was found in a generally good condition.

It was now possible to complete a clear passage of heading and tunnel and to walk through the tunnel the whole distance of 4 miles 628 yards. Progress was now such that by 8.0 a.m. on April 18th, 1885, the last length of brickwork was keyed in.

SEVERN TUNNEL (2)

Five months later (on September 5th, 1885) Sir Daniel Gooch (Chairman of the Great Western Railway) and Lady Gooch travelled by special train through the tunnel with a party of friends from Severn Tunnel Junction (Monmouthshire) to the Gloucestershire side and back.

Realising that the brickwork of the tunnel could not withstand the enormous pressure of the water, it was decided to instal an adequate permanent pumping plant to take away the water from the Great Spring, and an additional 29-feet diameter shaft and new pumping engines and pumps were provided at the Sudbrook and additional pumping plant at the Sea Wall Shaft.

This provision virtually completed the constructional work of the Tunnel.

∽   ∽   ∽   ∽

In order to make a rather difficult and much abridged story as clear as possible, I have made no mention of the machinery provided to supply air to the Tunnel and only brief references to the means of keeping water out of it, so we will conclude this talk with a short account of the ventilating and pumping plant installed.

When the Seven Feet Heading from side to side of the tunnel had been completed, a fan of 18 feet diameter by 7 feet deep was installed over the New Shaft at Sudbrook for the purpose of ensuring ventilation, air being drawn the full length of the heading. This fan of single inlet type was driven by a single non-compound engine, and a stand-by was provided on the opposite side of the crank pin, the two engines being worked alternatively coupled to the fan shaft.

Ventilating Fan, Severn Tunnel

The fan had an exhausting capacity of 70,000 to 100,000 cubic feet of air per minute, which was ample, for under normal conditions 30,000 to 40,000 cubic feet per minute sufficed.

On completion of the full-sized tunnel throughout, this fan was replaced by a permanent steam-driven one 40 feet in diameter and 12 feet wide, the installation being completed in August, 1886. This second fan was continuously at work for 38 years.

In 1924 the increase of traffic through the tunnel necessitated a more powerful installation, and the present plant was provided. The fan has a capacity of 800,000 cubic feet of air per minute, and is of 27 feet diameter by 9 feet wide, ample to allow for the fluctuating air currents caused by trains passing in opposite directions, a difficulty which, of course, is not met in ventilating a single track railway tunnel. The air is forced into the tunnel.

The driving engine is of the single tandem compound type. Three Lancashire boilers are provided, each 30 feet in length and 8 ft. 6 ins. in diameter, constructed for a steam pressure of 150 lbs. per square inch and fitted with superheaters for a temperature of 500-555 degrees Fahr. Two of the boilers are in commission at a time, the third acting as a stand-by.

As you have already heard, a good many springs of water were intercepted in driving the tunnel below the Severn estuary and in particular the Great Spring was broken into near Sudbrook. The necessary corollary was the provision of powerful pumping engines, temporarily at first, to keep the workings clear while the tunnel was under construction, and ultimately to deal with leakage into the tunnel after construction.

The pumping plant, as existing to-day, comprises :—

(1) The Sea Wall pumping machinery on the Gloucestershire shore ;

(2) The powerful Sudbrook pumping plant on the Monmouthshire shore ; and

(3) The plant at " Five miles four chains."

199

TRACK TOPICS

The main pumping station is at Sudbrook and you may like quite a brief account of the equipment there. The station is composed of three houses. No. 1 house contains single-acting Cornish beam engines with cylinders 70 in. diameter, 10 ft. stroke, which work six pumps—three bucket and three plunger type. The bucket pumps are 34 in. diameter 9 ft. stroke, and the plunger pumps are 35 in. diameter 9 ft. stroke; a pump of each type lifting 336 and 356 gallons per stroke respectively. These pumps deal with the Great Spring, the steam supply being generated by twelve Lancashire boilers 7 ft. diameter by 28 ft. long. There are usually nine in steam.

No. 2 house at Sudbrook has one 75 in. beam engine and two 50 in. " Bull " engines. The latter derive their name from the inventor. Both types are single-acting, taking steam on one side of the piston only. The 75 in. beam engine works a bucket pump 37 in. diameter 9 ft. stroke, lifting 400 gallons per stroke. The " Bull " engines have plunger pumps 26 in. diameter 10 ft. stroke, each lifting 218 gallons per stroke.

No. 3 house contains one engine, 70 in. diameter, working a plunger pump 37 in. diameter 10 ft. stroke, which raises 442 gallons per stroke. Nos. 2 and 3 houses cope with the drainage from the bottom of the tunnel, and the necessary steam supply is provided by twelve Cornish boilers, each 6 ft. diameter 28 ft. long.

The Sea Wall pumping station consists of two 41 in. beam engines working 29 in. pumps and the " Five miles four chains " house, which deals with drainage water from the Welsh end of the tunnel, contains two 65 in. beam

Severn Tunnel Pumping Station, Sudbrook

Boilers

Beam Pumping Engines

engines working one 34 in. bucket pump and one 35 in. plunger pump.

The height of the lifts in the respective pumping shafts is as follows :—

		Feet.
Sea Wall		106
Sudbrook, No. 1 house		180
„ Nos. 2 and 3 houses		200
"Five miles four chains house"		142

All the engines have continued at work as required since the opening of the Tunnel.

The maximum quantity of water pumped in any one day was 36,556,218 gallons, the minimum record being 13,374,332 gallons.

The first goods train passed through the tunnel on September 1st, 1886, and the works were reported as satisfactory by the Board of Trade on November 22nd, 1886. The first passenger train in regular running passed through the tunnel on December 1st, 1886.

The total length of the Severn Tunnel works, including the approach cuttings in the shore, land tunnels, and the portion under the river, is as near as may be seven miles, of which four and one-third miles are tunnels. Under the deepest part of the river the line is level for about twelve chains, but it rises with a gradient of 1 in 100 towards the English and 1 in 90 towards the Welsh side.

To-day we see the result of the thirteen years' fight by the indomitable engineers and their staff in a double line of railway line carried beneath the bed of the Severn estuary, along which an ever-increasing number of passenger, merchandise and mineral trains continuously passes on this direct route to and from the industrial centres of South Wales. The Tunnel also provides a direct route for Southern Ireland via Fishguard and Rosslare.

The Severn Tunnel, together with the South Wales direct line (Wootton Bassett to Filton), which was opened in 1903, reduced the distance to and from South Wales by no less than 25 miles, enabling the services to and from Wales to be considerably accelerated and improved and, as time is money, effecting substantial financial advantages to all users of the railway.

The quickest journey between Bristol and Cardiff in November, 1886, was made in two hours fourteen minutes, but the next month, when the Tunnel was open, the trip occupied eighty minutes. In 1881 the fastest train (the Boat Express of those days) left Paddington at 5.45 p.m. and took four hours twenty-three minutes to reach Cardiff and six hours to Swansea. To-day, the Fishguard Boat Express, which leaves Paddington at 7.55 p.m., takes two hours fifty-two minutes to reach Cardiff, and four hours to reach Swansea, whilst other express trains do the Paddington to Cardiff trip (with a stop at Newport) in two hours forty-one minutes.

$\backsim$ $\quad$ $\backsim$ $\quad$ $\backsim$ $\quad$ $\backsim$

I am afraid my story of the construction has been rather a recital of problems, difficulties, and disasters, but they were all faced with determination and eventually overcome. I do hope the account I have given you, if somewhat inadequate, has brought out the resourcefulness of those on whom the responsibilities of bringing the tunnel into being devolved, as well as the hardihood of the workers who fearlessly followed their callings under the bed of the river for so many years in darkness, danger, and discomfort, and frequently at great risk of their lives.

Reconstruction of Ponsanooth Viaduct

TALK NUMBER FIFTEEN

KEEPING THE TRACK LEVEL (2)
Viaducts

Y OU have already been told something in regard to construction of some of the early viaducts, and there is much in common between a viaduct and a bridge ; I think you will get to know more about them by taking a look at these photographs of some of the principal viaducts on the G.W.R. than from any verbal descriptions.

And here, in anticipation of a question you will probably ask, I am afraid I shall have to admit that I cannot explain why (for instance) we speak of the Royal Albert *Bridge* across the estuary of the Tamar, and of the Crumlin *Viaduct* across the Ebbw Vale. As a general rule, however (and there are exceptions to " prove " it), the term " viaduct " is employed when the structure spans something in the nature of a gorge or chasm, e.g., a Cornish valley such as Menheniot, Liskeard, and St. Pinnock. If this feature be lacking, the term " bridge " is more appropriate. The number of spans, by the way, has nothing to do with the choice of term.

The employment of viaducts for crossing valleys or gorges depends upon the height at which the railway has to be carried above the bottom of the valley, and the materials available for embankments. If chalk or rock can be easily obtained from adjacent cuttings, banks may be formed up to 60 feet, but if clay or other treacherous material only is available, it is usual to resort to the construction of a viaduct. It will be appreciated that the continuous maintenance of viaducts is much heavier than in the case of embankments. On the other hand, the maintenance costs of a viaduct are fairly well known, whereas in the case of an embankment of, say, 60 feet, several thousands of pounds might have to be spent on a single slip.

We have already seen that seventy years ago timber was extensively used in the construction of viaducts, such as those on the Plymouth to Truro lines and branches. With the exception of the Dare and Gamlyn Viaducts on the Dare Branch in South Wales, all Brunel's timber viaducts have now been replaced by more substantial structures of masonry or steel.

In some cases embankments have been substituted for viaducts, but that was only because suitable material was available, and the initial cost less, as well as that of future maintenance.

In the case of a brick or stone viaduct, there is generally little maintenance required for a number of years after construction, but where the superstructure of viaducts is of steel (and this also applies to bridges) painting has to be carried out every two, three, four, or five years, depending on the locality in which the structure is situated.

There are various methods of dealing with the maintenance of arch viaducts, the one most commonly used being a platform slung from the parapets and moved horizontally or vertically as required. Tubular scaffolding is now in general use, and more work can be done from the

Viaduct scaffolding

scaffolding. There must be some satisfaction to the workmen in having something solid to stand on when working at a height of one hundred feet or more above ground level.

Considerable developments have been made in the last few years in the method of painting structures such as viaducts, bridges, etc., by the use of compressed air spraying plant. This apparatus can be worked by electricity, petrol, or gas, and many types are capable of operating two or more sprayers together with a rotary wire brush, scaling hammers, and other tools for the cleaning of steelwork surfaces before spray-painting.

There are 92 viaducts on the Great Western Railway, of which Crumlin Viaduct is the longest. You will, doubtless, like to have a description of it.

We have to go to South Wales for what is, next to the Royal Albert Bridge, Saltash, probably the most important ironwork structure on the Great Western Railway. Crumlin Viaduct, near Crumlin High Level Station, spans the Ebbw Vale, through which runs the Western Valleys section of the Railway and the River Ebbw.

There are two sections—as a hill intervenes abruptly in the valley—and the larger is of seven spans of 150 feet each, supported on six piers; the smaller is of three spans of 150 feet each, supported on two piers. The greatest height is 193½ feet from the rails to the average level of the River Ebbw, and 208 feet from the rails to the bottom of the pier foundations.

This photograph gives you some idea of the dimensions of the viaduct. The piers are of identical construction, but varying in height according to the contour of the

Crumlin Viaduct

valley. Each one consists of fourteen cast-iron columns, stayed at intervals by cast-iron distance pieces and braced by wrought-iron ties.

Each span of the superstructure consists of four main girders and originally the decking was of timber laid transversely over these. In 1865 it was found necessary to strengthen the superstructure, and at this time an iron deck was substituted. The main girders are of the inverted " warren " type, the top or compression members being box-sided in cross sections. The bottom, or tension members, are of flat plates on edge. Those web members which act only as tension members are also of flat bars, but such as have to resist compression are cruciform in cross section.

The viaduct, which cost £62,000, or £41 7s. 0d. per foot run, was designed by Mr. Thomas W. Kennard and erected under his personal supervision. An interesting feature of the construction was that, owing to the design

Cefn Viaduct

adopted, the use of expensive scaffolding was obviated as each tier of columns, as erected, served as scaffold for the next above, and by omitting some of the cross bracing until the end of the work, spaces were left in the piers by which the girders, put together on the ground, were hoisted by means of temporary timber crossheads into position at the tops of the piers. The raising of each 150-feet girder span, weighing with its temporary bracing 25 tons, was a day's work for twenty men with block and crab tackle.

For the foundations, the ground was excavated to firm gravel and a foot of concrete laid to form a bed for a solid mass of masonry, twelve feet high, capped by stone blocks a foot thick to which the base plates of the columns were secured by twelve one-inch diameter bolts.

The work of construction was completed on June 1st, 1857, three and a half years after it commenced. It was tested by six locomotives loaded on every available space with pig iron and rails and, with all the engines on a span, the greatest deflection was one and a quarter inches.

VIADUCTS

You will probably wonder what the effect of heat expansion is on such a long all-metal structure. Some trouble was experienced from expansion and contraction, and a slight alteration had to be made by supporting the girders at the abutments on rocking links. The greatest variation caused by differences in temperature in one day was seven-sixteenths of an inch on the total length of the viaduct, and that was between 4.0 a.m. and 4.0 p.m. when a fine June day followed a cold night. The greatest differences in length recorded aggregated two and a half inches, between temperatures of 32° and 90° Fahr.—a difference of 58°.

The viaduct carried a double line of rails until 1927, when it was converted to a single line structure in order that two heavy mineral engines might work over it coupled.

The quantity of metal used in construction of Crumlin Viaduct is 1,300 tons of wrought iron and 1,250 tons of cast iron.

~   ~   ~   ~

So much for viaducts. Before coming to the subject of bridges, I want to have quite a short talk on gradients.

Coldrennick Viaduct

Diagram of Gradients

Paddington to Penzance via Westbury

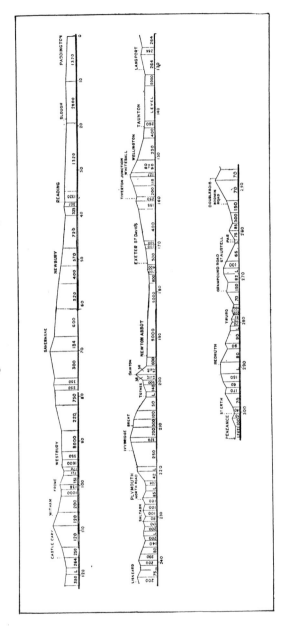

TALK NUMBER SIXTEEN

GRADIENTS

WHAT is a gradient? Well, there are some loosely-worded definitions, but I think " *a gradient is the proportion which the vertical rise of anything bears to the horizontal distance travelled* " is good enough for our purpose. As you already know, variations in the rail level are denoted by the number of lineal feet in which the track rises or falls one foot, and these changes in the level are indicated by gradient boards fixed at the side of the line.

Let's take a look at the gradients on the route of the Cornish Riviera Express, that is the direct line London to Penzance via Westbury. As we have seen, on the old route via Bristol, Brunel kept his track practically level as far as Taunton, and almost dead level to Swindon. The direct route to the West joins the old route via Bristol at Cogload Junction near Taunton.

The direct route from London to the West leaves the old main line at Reading, and here is a diagram of the gradients by that route, Paddington to Penzance. Take a good look at it, for it is interesting.

As you see, there is no gradient worthy of the name between Paddington and Reading. The line then rises 1/307 and falls again 1/323 by the 40th milepost. From this point it rises steadily through Newbury all the way to Savernake with a sharper rise (from Bedwyn) to the Savernake summit of 1/184.

The line then is generally on a falling gradient to beyond Westbury, and rises to Frome to drop again and then rise steadily to a point just beyond Witham, when it falls rather sharply. Then, after a short piece of level, it rises at Somerton and then drops again.

Before reaching Cogload Junction, where it is joined by the old route from Bristol, there is a stretch of dead level for nearly ten miles.

Between Taunton and Wellington the line rises on an average of 1/400, then at Norton Fitzwarren it steepens to 1/230, the last four miles from Wellington to Whiteball being 1/80 to 1/90 with a short length of 1/127 to the summit, from which point the line falls generally to Exeter and is practically level from there to Newton Abbot.

As you see by the diagram, things change drastically after Newton Abbot, and some of these gradients are a legacy (as you have heard) of the old South Devon Railway, laid out originally for atmospheric traction.

Look at that " peak " at Dainton. The rise includes a gradient of 1/36 while the fall is one of 1/37, and then after Totnes comes the Rattery Bank (Brent), the first part of which has gradients of 1/50 and 1/54 easing to 1/100 and then 1/500. Then there is Wrangaton ahead (near Ivybridge), with a gradient to the summit of 1/200, where the line is 462 feet above Ordnance datum. From this point (15 miles from Plymouth) the line falls to within a short distance of North Road Station, where a stiffish climb of 1/85 occurs.

From Newton Abbot to Plymouth, there is some hard going (and also in the reverse direction) ; in fact, this route has some of the hardest gradients of any main line used by express trains in Great Britain. Fortunately, in the matter of load, speed and reliabilty, G.W.R. locomotives are in a class of their own and the famous " King " engines and others can tackle these gradients comfortably with quite heavy trains.

The diagram of the line Plymouth to Penzance looks something like a switchback, and there is indeed very little level road on the whole 80 miles. The gradients from Plymouth to Liskeard include two short lengths of 1/80 and one of 1/75, and there is a fall beyond Liskeard of 1/200 rising again at 1/70 to Doublebois, and falling in same ratio ; and then at 1/150 and 1/300 after a rise (1/85) and a fall (1/75) just before Par Station, where the level of the line reaches its lowest point.

We then get a steady climb of 1/65 to St. Austell and onwards at 1/100. The line then falls to rise again at Grampound Road, and drops from there at 1/70 and then, with variable gradients, over the five miles to Truro.

The highest point between Truro and Penzance is Redruth, reached by gradients of 1/70, level, and 1/90 (a short fall at 1/90) and then 1/80. From Redruth the line falls generally (there are only two short up-grades) to Penzance.

Well, that gives you some idea of what the G.W.R. locomotives have to face in the way of gradients between Paddington and Penzance, and you have only to follow the diagram in the opposite direction to see that the east-bound (up) trains have also to do some pretty heavy collar work, for, of course, what is an easy run on a falling gradient for a down train, becomes a stiff climb for an up train and vice versa.

TRACK TOPICS

In our chat on the Severn Tunnel I mentioned the South Wales and Bristol direct Line (from Wootton Bassett to Filton) and I should like to refer to it again here, for it is an example of " keeping the track level " from the railway engineer's point of view. The line is so well constructed that, although it traverses hilly country, there is no gradient steeper than 1/300 and no curve sharper than 1 mile radius. Throughout its whole length, however, the line is nearly evenly divided between embankments and cuttings, and its construction involved the excavation of five million cubic yards of material, with a slightly larger quantity tipped for embankments. But that is not the full tale by any means, for four viaducts had to be built, 100 bridges, and two tunnels, and of the latter one—under the Cotswold Hills —is no less than two and a-half miles in length !

The two highest altitudes reached by the rails on the G.W.R. system are at Princetown in Devonshire and at Dowlais Top in South Wales.

You may know Princetown, which is the terminus station of a branch from Yelverton on the Tavistock Line, on the western side of Dartmoor. On this branch the line rises 850 feet in $10\frac{1}{2}$ miles, the steepest gradient being one in forty, and the average for the whole distance one in sixty-five. The line at Princetown is 1,373 feet above sea level. Incidentally, it is quite near the big Dartmoor prison, but I don't suggest that your acquaintance with Princetown may be in that connection.

Dowlais Top is the summit of a steep incline of 1/40 in the mountainous coal producing area of South Wales and is 1,314 feet above sea level. A story is told of a heavy freight train toiling up this incline, and the driver

Princetown, the highest point on the G.W.R.

of the locomotive having to make his engine go " all-out " to bring his train to the summit, whereupon he mopped his brow and remarked to his fireman (who must have been a new hand, and surely from the Emerald Isle) that it was one of the stiffest climbs he had ever experienced. To this his fireman mate remarked, "Yes, bedad, an' if I hadn't put the brakes hard on we should have been slippin' back all the way." Well, that's the story, quite libellous I expect, and I really can't imagine it happening—well, not on our Great Western Railway.

You would naturally expect to find some fairly stiff gradients in Wales and you would not be disappointed. On the Pwllyrhebog branch, which runs from Tonypandy and Trealaw in the Rhondda Valley to collieries high up on the mountainside, the line rises on a gradient of one in thirteen for about half-a-mile, and on this portion the engine and train of empty mineral wagons going up the gradient are assisted by means of an endless steel rope. The loaded coal wagons coming down the gradient are also attached to the rope, and their weight provides the extra hauling necessary of the up-going train. At the top

Cable working on incline

of the one in thirteen gradient is a goods yard, and from here the line continues at a varying rise from 1/29 to 1/30 for a distance of about one mile.

In the Llanelly district we get another of these mountain branch mineral lines worked by wire rope, and in this case the gradient is as stiff as 1/12.

On the Bala-Festiniog Line there is an average gradient of 1/55 for ten miles, rising from Bala to the summit. The line then falls to Trawsfynydd for a length of seven miles from the summit at an average gradient of 1/58.

On the Dare Branch the passenger train service terminates at Aberdare, but the line continues as a mineral railway for about two and a half miles and the greater part of this extension is on a gradient varying from 1/38 to 1/30.

TALK NUMBER SEVENTEEN

BRIDGES

As there are nearly 12,000* bridges of various dimensions on the Great Western Railway, you will realise that their construction, maintenance, and renewal present many problems for the engineers, particularly for the steelwork section, which is responsible for the designing and details of construction of all bridges wholly or mainly of steel.

In the matter of material employed, brick or stone is the first choice for bridges, depending upon the locality of the structure. Steel superstructures are built in situations where brick would be impracticable owing to span, angle of skew or other physical feature, and sometimes on account of lower first cost.

Ferro-concrete has been used on occasions for bridges over the railway, but seldom, I think, for carrying a railway track.

Design is necessarily based upon traffic requirements and the physical features of the site, certain stipulations as to strength being prescribed by the Minister of Transport.

The longest bridges on the Great Western Railway are Barmouth Estuary, 2,292 feet, and Saltash (Royal Albert), 2,200 feet. The two main spans of the latter, as you already know, are each 455 feet in length, and these are the longest bridge spans on the G.W.R. system.

*Exclusive of dock system (245) and canals (460).

Chepstow Bridge

The largest arch bridges are over the Severn at Arley and Coalbrookdale. These have spans of 205 and 201 feet respectively, and are of cast-iron.

The longest arches, not of metal, are the two of Brunel's bridge spanning the Thames at Maidenhead. These, as you already know, are 128 feet clear. By the way, you may be rather surprised, I think, to learn that the G.W.R. track crosses Old Father Thames at no fewer than ten points.

There are several remarkable bridge structures, some even unique, on the Great Western Railway, and we have already considered one of them.

    ᔕ   ᔕ   ᔕ   ᔕ

I think the River Wye Bridge at Chepstow calls for special mention. This is one of Brunel's works, built to carry the double lines of the South Wales Railway, and like the builder this bridge is unconventional. Its chief feature is its superstructure, as you see from this photograph.

The bridge is of four spans, one of 300 feet, which almost spans the width of the river, and three of 100 feet each at the western end.

It carries two lines of way, each on a separate bridge, with a space of four feet between the two bridges. The 100 foot spans are of plate main girders of ordinary construction seven feet deep, and girders of similar proportions, but assisted by a rather unique trussing arrangement, are employed over the main span of 300 feet.

Centrally over each track at about 50 feet above rail level there is a wrought-iron tube over 300 feet long and of nine feet diameter, made of wrought-iron plates, one end of which is supported on a tower carried by the cylinder sub-structure and the other end on the solid rock as you see in this photograph. The trussing is formed by the connection of the tube and the plate girders by means of two substantial vertical frames and flat bar ties, the whole forming an inverted " queen-post " arrangement.

This bridge was opened for traffic in 1852, and was Brunel's first large iron bridge—the construction of Saltash bridge was not commenced until two years later. In conception it was a considerable advance upon the practice of the time as evidenced in the work of Stephenson and Fairbairn at Menai Straits and elsewhere.

The greatest height from high water level to the top of the large curved tubes is over 100 feet. The head-way under the span is 50 feet at high water, to suit the needs of vessels navigating the river. The total weight of cast-

The Swing Bridge, Carmarthen

iron in the structure is 2,340 tons, and of wrought-iron 1,278 tons.

∽   ∽   ∽   ∽

When a bridge carrying the railway crosses a navigable river or estuary, it is in some cases necessary to provide for a portion of the bridge to open so as to afford unlimited headway for vessels using the waterway.

Opening bridges are of several types : the best known, perhaps, is the " swing " bridge, which moves on a centre pivot so that in its open position the swing-section is at right-angles to the railway. An example of this is found in the Barmouth Estuary Bridge.

Another form is the " bascule " bridge, in which the outer end of the movable span rises vertically : the Tower Bridge, London, is a famous example and this is a " trunnion-bascule," the opening span rotating about a fixed shaft or trunnion. A more modern form is the " rolling-lift bascule " and this principle has been employed in the bridge which carries the South Wales Main Line over the

Towy Estuary just below Carmarthen town. Such a bridge does not pivot about a shaft as the " trunnion-bascule " does : it *rolls* backwards, lifting simultaneously.

The bridge consists of five steel fixed spans and one opening span, giving 50 feet of clear waterway when open.

Here is a diagram showing the operating gear of the lifting span which, together with the accompanying

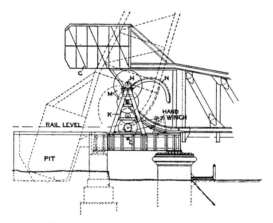

Diagram shewing operating gear to the
Lifting Span of Carmarthen Bridge

photographs of the lifting span closed and open, demonstrates the principle. The main girders of the opening span are lattice type, tapering eastward on top to meet the level of parapet to the fixed spans, the rail end being made of heavy plate box-section with the bottom flange curved to form a quadrant giving sufficient roll when opening for the 50-feet clear waterway. The operation of opening (lifting) and closing the bridge is facilitated by the provision of balance boxes (G), which are fixed on the top of the bascule girders. These are filled with cast-iron blocks run in with asphalt.

In lifting and closing the bridge a curved cast-steel rack (N) is provided on the outside of each bascule girder, into which a cast-iron pinion (H) at the rolling centre of the bridge turns. As will be seen, pits are provided for the balance boxes when the opening section is lifted.

The power exerted for lifting and closing is worked through differential spur-wheel-gearing (M), supported on axles and bearings attached to a steel braced trestle (K) on to a connecting shaft (L) extending across the width of the bridge. The trestles are fixed at the base to cantilevers attached to the bearing girders.

The direct power exerted to the connecting shaft is supplied by electric motors, and there are also electric brakes for automatically controlling the bridge against any acceleration when lifting or closing, and also for locking the opening span when open. The motor and brakes are driven by a battery of accumulators, charged by an oil engine coupled to a dynamo. There is also a hand-driven winch provided for operating the bridge in cases of emergency.

For locking the bridge when closed for traffic a steel bolt is inserted on the cross-girder of the fixed span, which moves into a cast-iron slot connected with the cross-girder of the opening span. This bolt is worked by the signalman in the box near the bridge, the lever being connected with the electric controller, thus making it impossible to open the bridge while the bolt is in the slot or while the signals are off.

BRIDGES

The parapet railing beyond the bascule girders at the west end is collapsible, in order to clear the bracing girders between the balance boxes when the bridge is opening. This railing is worked by the machinery in order that it may drop by the time the balance boxes are clear of the pits. The pits and abutment are built of brickwork in cement, on concrete foundations, the foundation to the abutment being reinforced with old rails and supported throughout by timber piles.

The bridge takes about three minutes to open or close.

෴   ෴   ෴   ෴

The Barmouth Estuary Bridge (equally well known as Barmouth Viaduct) was inherited by the Great Western Railway when it took over the Cambrian Railway, under the Railways Act of 1921. The bridge is mainly a wooden structure of 113 spans, but on the north end there is a fine steel-work section, including a swing portion which when open provides two passages for vessels to pass from the open sea to the Estuary. This section is supported by cylinders which had to be taken down to the bed of the estuary to reach a rock foundation. There is a footpath for pedestrians across the whole of the bridge.

Barmouth estuary bridge is well known to the many thousands of holiday-makers who annually visit the lovely Cambrian Coast, for the estuary which it spans is one of surpassing beauty. At any time of day, but particularly, perhaps, in the " westering sun," the long bridge strung across the blue waters of the estuary, the golden sands, and panorama of surrounding hills, unite to make a scene which is surely matchless for its colourful magnificence.

225

Barmouth Estuary Bridge

Looking from the Bridge towards Dolgelley, Cader Idris is the background of a really glorious scene.

But to-day Barmouth Bridge is familiar to a wider public than those who have actually visited the locality, for it has an extensive film reputation. Perhaps you can be counted among the myriads of spell-bound spectators who have been thrilled by the story of the "Ghost Train."

All the railway scenes (other than studio sets) for that popular film were "shot" at various places on the Great Western Railway, and you doubtless recall the sensational climax in which the Ghost Train with its nefarious freight plunges over the opened bridge to its watery grave.

There is much in that story, by the way, which is quite "*un*-railway" in practice, but that does not matter perhaps in a stage or film plot, so long as it provides a good "story"; even when the detective-comedian-hero rushes into the signal box and, aided by a charming lady, operates the lever opening the swing bridge over the estuary and so sends the train to its doom.

The stirring incident of the Ghost Train's destruction was photographed at Barmouth Bridge, and was, of course,

By courtesy of Gaumont British Pictures Corporation

FILMING "THE GHOST TRAIN"

Above : Model train pitching from the Bridge.

Left : Signal Box Scene being filmed.

Below : Model bridge erected for the film with the actual bridge in background (reflected in mirror).

achieved by a piece of trick photography, but it was very cleverly contrived. After a train had been photographed passing over the closed bridge at Barmouth, the " shot " in which the Ghost Train was running on to the open swing bridge was taken by means of a mirror, which obscured the actual bridge and a short piece of track (but not the surrounding scenery) and into this mirror was reflected an exact model of the bridge with the swing section *open ;* the model being erected near the cameras.

The Ghost Train was filmed travelling towards the bridge and at the appropriate moment a working model train—a replica of the real Ghost Train—was set in motion on the model bridge and timed to take up the running of the real train (as it disappeared from view, and to safety behind the mirror), and the model train was photographed pitching headlong from the bridge down to the waters of the estuary, where it was engulfed. A nice, tidy end for the Ghost Train.

Well, that's how *that* little bit of deception was managed.

ﺹ　　　ﺹ　　　ﺹ　　　ﺹ

The ordinary method of sinking cylinders as foundations for bridges is to lower the cylinder on to the ground, excavate and grab out the material from inside, and then by loading the top of the cylinder with heavy weights—usually in railway work old rails—the cylinder gradually sinks. The bottom of the cylinder is provided with a cutting edge, and as the material is removed from inside, the cylinder cuts its way into the ground.

A recent innovation consists of blowing out the material from inside the cylinder by means of compressed air.

BRIDGES

The bottom of the cylinder is fitted with a cutting edge having a series of holes on the inside, which are connected by a pipe to an air compressor. The cylinder is first driven into the ground a short distance, the air pressure is then turned on, and this forces the piece of ground inside the cylinder upwards and out through the top of the cylinder.

From time to time we read in the daily press an account of the replacement of a railway bridge, and much is made of the fact that this feat is accomplished without disturbing the normal traffic.

Judged from the size and " snappiness " of the head-lines, this sort of job is considered in Fleet Street as something worthy for a generous flow of superlatives, and we read a " story " in which modern journalese has full play. " Mammoth " cranes swing " gigantic girders " about and all is hustle and staccato activity checked on a stop watch by the heroic engineer who, working to split seconds, has to get the gap bridged before the midnight mail is due—or something of that sort. All very fine, but not much like the real thing, I am afraid.

All railwaymen and others know that bridge replacements are among the ordinary business of the railway engineers, and hardly a Sunday passes but some such job —and possibly several—is carried out. What really controls the amount of work done on a Sunday, or over a week-end, is the preparatory work which may have taken weeks, or even months.

The principal reasons for the replacement of bridges are loss of material due to corrosion, the intention to run heavier engines, and the widening of public roads.

TRACK TOPICS

The G.W.R. will rebuild about 130 bridges during 1935. Thirty-four, already completed, are situated between Bala and Trawsfynydd, North Wales, and the principal reason for this work was the desire to run more powerful engines with appreciable saving in running costs.

∽   ∽   ∽   ∽

Some idea of the factor of safety provided in modern steel bridges may be gathered when it is realised that, in general, the stress produced by the heaviest engines does not exceed 8 tons to the square inch of metal, of which the known breaking strength is not less than 28 tons to the square inch. Similar margins are provided in the other materials used in bridge-building.

All questions of design and stress are dealt with in the office of the Chief Engineer, and in carrying out bridge reconstruction work there has, of necessity, to be the closest collaboration between the Engineering and the Traffic Departments. By the use of modern equipment and methods, and by having much preliminary work carried out before obtaining " occupation " of one, or both roads, from the Traffic Department, interference with the working of trains is reduced to a minimum.

Cranes, capable of lifting up to 36 tons each (used in pairs for handling long girders), hydraulic lifting jacks, pneumatic riveting plant, and oxy-acetylene cutting apparatus all play a part in modern bridge reconstruction.

In cases of reconstruction where the superstructural work is of steel any necessary work on the sub-structure is usually carried out while the several pieces of steelwork are being fabricated.

BRIDGES

The nature of the principal operations at the site differ for many reasons. Where the physical features permit, the complete new bridge may be built alongside the old one and the exchange effected, during an occupation of a few hours, by the process of rolling out the old and rolling in the new, this of course necessitating some kind of temporary work to act as track for the moving masses. Generally, and in the case of the smaller bridges, the complete renewal is made piecemeal by cranes working at either end of the span. When the span is considerable it may be necessary for the cranes to lift heavy loads while standing on the obsolescent work and this may require local strengthening for such operations. Such a case was that of the Eign Bridge over the River Wye.

The old bridge consisted of three spans of approximately 43 ft. 6 in., 73 ft. and 43 ft. 6 in., and carried two lines of track. The new structure, which is of the same overall length, was built in two spans by the provision of a new central pier. This consisted of two steel cylinders surmounted by a cross-head girder on which the new main girders were seated.

In reconstructing bridges of similar dimensions it is usually found necessary to resort to single line working at the site for some time, which necessitates the provision of running junctions on either side of the bridge and additional signalling. In this case there was convenient siding accommodation available near by, and this was utilised for assembling the whole of the new superstructure which would have to carry the down line. This consisted of two spans of 83 ft. 9 in. each overall, weighing 86 tons. These spans were riveted up, and bitumastic floor covering

Destroying the old Eign Bridge

was laid complete, ready for the reception of the ballast and permanent way.

An occupation of both roads at the bridge site extending over a long week-end was then arranged with the Traffic Department. This commenced at eleven o'clock on a Saturday night, and the demolition of the old super-structure carrying the down line was pushed forward vigorously. At the same time two 36-ton cranes were engaged at the siding in loading up the two complete spans. One of them was loaded on a G.W.R. 120-ton

One 86-ton span on " crocodile " and the other being loaded

" crocodile," and the other on an 80-ton vehicle. These were exceptional " out-of-gauge " loads, the maximum height above rail being 18 ft. 9 in., and the overall width 15 ft. 8 in.

By noon the following day the old superstructure carrying the down line had been removed and loaded up for removal, leaving only the old cast-iron columns in the river to be dealt with. These had to be got out of the way before the new superstructure could be placed in position. They were dealt with by tipping them over into the river bed, whence they were later recovered.

The columns offered considerable resistance to overturning. Each required a pull of nearly 100 tons by a steel hawser, one end of which had been attached to the head of the cast-iron column and the other to a winding engine anchored in a neighbouring field.

By ten o'clock on the Sunday night the two new spans had been placed in position, and four hours later, at two o'clock on Monday morning, the new permanent way and ballast laid, and the down line ready for traffic.

During a similar week-end occupation of the two lines on the following Sunday, the remainder of the old bridge, carrying the Up line, was removed and the new superstructure substituted piecemeal.

You may be interested to know something of another method of bridge demolition. Occasionally a bridge has to be dealt with by the drastic process of blowing it up. By the explosion of twenty-seven pounds of gelignite an arch bridge of 68-ft. span, near Bristol, was lately demolished. The structure was an old brickwork one, 13 ft. wide and 3 ft. 3 in. deep throughout.

Ready

Going

Gone !

BRIDGES

Prior to the demolition the earth and clay, together with the spandrill (side) walls, were cleared away, down to the extrados (top) of the arch. Then fifty-four charges were laid. Each consisted of two 4-oz. cartridges of gelignite (making a total of 27 lb.). The charges were arranged in seven rows and drilled to an average depth of 2 ft. 3 in. They were placed not less than two feet from the edge of the arch and not more than two feet apart.

To prevent damage to surrounding buildings torpedo netting, timbers and brushwood were piled over each row of charges, and the windows of the booking office of a halt, 4c feet from the bridge, were boarded over. Two lengths of rails and sleepers were removed from each line of track immediately under the bridge.

If an operation of this kind is to be successful a good explosion is essential and great care necessary in selection of detonators. In this instance the charges were fired by means of two submarine electric detonators, one being connected to thirty-seven of the charges, and the other to seventeen. The explosions were simultaneous, and the result quite successful.

The arch dropped clear at springing level and the debris was well broken up. The explosion took place at a quarter past nine in the morning. A six-ton steam travelling crane, with engine attached, moved up immediately afterwards, and the clearing of the debris was taken in hand. The up road was connected up at one o'clock, and the down road two hours later, and that was *that*. Pretty smart work, eh?

Here are some photographs of the bridge before, during, and (what *was* the bridge) after the explosion.

An exceptional load of Buoys

TALK NUMBER EIGHTEEN

THE LOAD GAUGE
AND EXCEPTIONAL LOADS

Y ou have probably seen the standard load gauges which are erected at practically all railway stations and points of exchange. That appliance is a convenient means of ascertaining if a loaded wagon is within gauge and safe for travel—as to height. But height is not the only dimension of which account has to be taken.

Take a look at this load gauge diagram. It represents the shape and size of the largest load that can travel in safety on the G.W.R. system in ordinary working. It does not, as has been stated more than once, represent " the lowest arch which trains have to pass under anywhere on the line," for a margin of several inches, at least, is necessary between a load and any arch or structure it passes to allow for lateral and vertical oscillation.

The diagram is used for traffic purposes, and loads of greater dimensions than those indicated are only allowed to travel by special arrangement—as " exceptional " loads. You will notice that in this diagram there are no clearance dimensions between rail level and a horizontal line 3 ft. 6 in. above, for no load may exceed the width of the wagon on which it is carried.

The height of the G.W.R. standard load gauge is 13 ft. 6 in. and its full width 9 ft. 8 in., and this latter dimension is greater than that of any other railway in Great Britain.

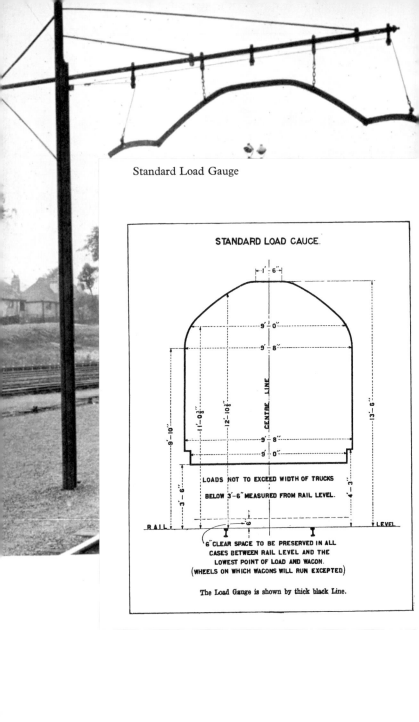

Standard Load Gauge

The Load Gauge is shown by thick black Line.

THE LOAD GAUGE

Now, if you look at the diagram again, you will see that the height of 13 ft. 6 in. extends only for 9 inches on each side of the centre line and that the top of the gauge consists of curves of different although definite radii with 18 inches (i.e., 9 inches on each side of centre) of straight at the summit.

Then again, you will notice that the extreme width is only available between 4 ft. 3 in. and 9 ft. 10 in. above rail level, while below the 4 ft. 3 in. plane the width is 9 feet only, being further diminished below platform level.

But there is another important dimension, which cannot be shown in the diagram, and that is length. You have only to visualise a load such as a long baulk of timber, steel mast or girder loaded on two or more wagons to realise that while passing over a curve (though the wagons would conform approximately to the arc of the track), the long load, owing to its rigid character and allowance for lateral play, would want more clearance. Now substitute for your rigid mast or girder a load of the same length, but of full gauge width, and you will at once see there would be a considerable overlapping of the gauge.

It is for that reason that the load gauge is limited to loads not exceeding 40 feet, and with loads of greater length the width permissible is reduced.

Loads of 40 feet but not exceeding 50 feet in length must not be more than 9 feet wide, and

Loads exceeding 50 feet but not more than 60 feet in length must not exceed 5 feet in width, while

Loads beyond 60 feet in length are " exceptional " loads and can only be carried as such, i.e., by special authority in each case ·

Just so, but you will probably ask " What about the G.W.R. 70-feet coaches themselves ? " These, of course, have not the variability of lateral movement that has to be accommodated with goods traffic, which comes freely forward on the wagons of all railways, with widely different spacing of wheels and bogies and endless variations ensuing therefrom in the amount of " side play " as well as even greater diversity in character of the loads. These 70-feet coaches, I may tell you, are subject to many limitations in running.

With coaches, a trial trip with one of the type is conclusive, especially as any little improvements of the clearance which are desirable are seen and noted and can be made before the coach is allowed to work in ordinary traffic conditions. The same situation may be taken to hold good for locomotives, many of which are built out to the full gauge limits in some respects, especially the cylinders.

There is another factor, too, closely involved with that of long loads travelling round curves. You already know that on curves the outer rail is raised above the level of the inner, known as the " cant " or " super-elevation " of the track, the object of which is to enable the curve to be taken at some speed in safety. Now, as you will readily understand, the effect of raising the outer rail is that engines, coaches and wagons are canted over—and it is a factor of safe transit that they should be—on the inner side of the curve. Consequently, when a long load is projected outside the gauge in rounding a curve, by the lateral " play " over the bogies, that projection may be substantially increased by the tilt due to the cant, especially at the top should the load be a high one.

THE LOAD GAUGE

Now I am sure a moment's thought will indicate why the load gauge you see at ſtations is always hung over a *straight* length of line. If the gauge was placed on a curve, loads passing under it for teſt would, due to the combined effeĉts of curve and cant, occupy quite eccentric positions relative to the gauge.

Perhaps I ought to make it clear that special clearances are always provided beside curves of short radii. Here is a diagram which illuſtrates my point. It shows two long coaches passing each other on adjoining lines on a sharp curve. As you see, the advancing corner of the coach on the inner road is being projeĉted towards the outside of the curve, simultaneously with the projeĉtion of the mid-length of the outer coach towards the inside of the curve, thus absorbing more than might be anticipated of the clearance of the " six-foot way."

Before we leave this subjeĉt there is yet another clearance which ought to be mentioned. A clear space of six inches above rail level is required by the load gauge to be preserved *beneath* the bottom of all loads.

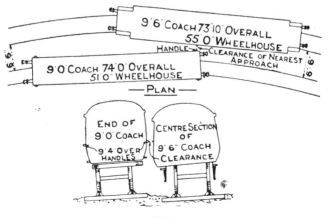

This six inches of space muſt not be encroached upon by any part of a locomotive, passenger coach, wagon or load ; allowing, of course, in the case of a locomotive for the loweſt position of moving parts. It is true, however, that one or two projeĉtions into this six inches are recognised in conneĉtion with cab signalling and eleĉtrical working, but they can be dismissed with a mere reference.

I might add that the ſtandard gauge does not apply to a few branch lines, sidings and goods sheds where ſtruĉtures impose a rather less generous minimum.

ᔕ   ᔕ   ᔕ   ᔕ

The load gauge leads naturally to the subjeĉt of " exceptional " loads, that is, loads outside the ſtandard gauge, which are only conveyed by special arrangement to ensure the special clearances required.

An " exceptional " load is one which exceeds the load gauge dimensions in height, width, or length, or in width as affeĉted by length on curved track. Incidentally, it may also be exceptional because of weight, but that is not exaĉtly a load gauge queſtion.

I don't think I need bother you with the whole procedure of dealing with exceptional loads, except to remark that when such loads offer, the firſt thing done is to seleĉt the moſt suitable wagon (according to the load's dimensions) and ascertain from the Engineering Department if by one route, or possibly an alternative, the clearances between the points concerned permit of the load passing safely.

The great majority of exceptional loads exceed the gauge in width, and loads as wide as 10 ft. 4 in. are not uncommon. Sometimes they run to greater width. Such consignments may be boilers, vats or tanks, ships' pro-

The Casing for Europe's largest Electric Generator—an exceptional load

pellers, big and heavy electrical plant, or coach bodies or underframes consigned to ports for shipment abroad, notably to South America and Egypt. These often travel over considerable sections of line as, starting in the Midlands, their place of shipment—dependent upon the vessels available—may be a South Wales port, Liverpool, or London. Indeed, the relative frequency of such loads has led to a special set of instructions and schedules being issued, detailing the places where trains conveying a special load may pass another train on an adjoining line and where sidings next to the line in use must be cleared. These schedules are based on the width of clearance, and are more numerous as the loads provided for grow wider.

Exceptional loads, as you can well imagine, are of various kinds and various dimensions, and as such require special arrangements to be made for their conveyance. We cannot consider them all, but I may add that it is often necessary to run such loads under very close restrictions ; for example, to require adjoining lines on both sides to be kept absolutely clear of traffic, a point of importance being the lateral " play " of long and wide loads on bogie vehicles when passing curves. Speed often has to be limited to walking pace at tight bridges, etc.

Passenger coaches manufactured in England for foreign and colonial railways are usually much wider than our maximum loading gauge. This, combined with their length, often presents pretty problems in transit from Works to port. The difficulty with excessive width has sometimes been met by raising the bodies from the bogies about eighteen inches. In some instances where conveyance is particularly difficult a more ingenious method has been devised. The coaches are carried on special bogies fitted with apparatus by which they can be traversed from side to side as much as eighteen inches when passing platforms, signals, arch bridges, girders between tracks, etc.

As you will have already observed, the variety of special wagons in service on the Great Western Railway has done much to facilitate the handling of loads of unusual dimensions. The character of the " pollen " or girder wagons will be appreciated from this photograph in which are seen two long girders. The pollens are twin trucks with turntables in the floor of each. The loads are borne wholly on these turntables and secured thereto also ; with the effect that the trucks practically serve as the bogie undercarriages of passenger coaches, the girders, to complete the simile, standing for the coach bodies.

Girders weighing together 138 tons for conveyance by rail

We cannot even leave this subject, particularly in view of the growth in the number of exceptional loads offered in these days, without another testimony to the foresight of the great Brunel, whose broad gauge load gauge was 15 ft. 6 ins. in height and 11 ft. 6 ins. in width. It would certainly have taken a good many of the exceptional loads offered to-day well out of the " exceptional " category.

You may be interested to know that for engine working the lines are classified on maps into seven classes—" hatched red," " red," " dotted red," " blue," " dotted blue," " yellow," and uncoloured.

As a G.W.R. locomotive enthusiast you have noticed a small coloured disc in the side of each locomotive cab. These discs correspond with the colours on the engine working map and indicate the lines over which locomotives of varying types and weights can travel. For instance, " the ' Kings' ' highway "—the lines on which the " King " class engines may run—is " hatched red," while on lines coloured " red " or " dotted red " all classes of locomotives *except* " Kings " may run—and so forth.

New façade, High Street Station, Swansea

TALK NUMBER NINETEEN

RAILWAY BUILDINGS

As we are primarily concerned with the railway track you will not expect me to say much about station and other buildings, although their construction and maintenance are matters for the Engineering Department. In method of construction station buildings are much like other buildings, but not quite like (shall we say?) some of the modern " council houses," for in order to withstand the vibration of trains they have necessarily to be much more substantially built.

I have here photographs of stations, some of which have been taken while under reconstruction, which you will, I think, find of interest, and you will also appreciate that the rebuilding of such stations as Cardiff, Newport, Swansea, and Bristol, to quote recent cases, with little or no interference with the ordinary traffic working, particularly through the busy summer months, calls for considerable forethought and organisation.

In civil engineering work on railways one of the outstanding features of recent years is the increasing use of cement concrete in various forms. Not only are many structures previously built of steel now made of reinforced concrete, but concrete in novel forms is being used in place of timber, brickwork and masonry.

You have probably noticed the concrete fencing posts alongside the railway. These have replaced timber and are both stronger and more lasting. Incidentally, you may not know that Great Britain is the only country in which the railways have to be fenced off in this way.

Reinforced concrete can be seen in much of the modern building work, but a case peculiarly applicable to railway work is that of the pre-cast concrete platform wall. This design was used in the platform extensions at Paddington and several other recent works. Instead of constructing the platform wall of brickwork, it is formed by placing in position sections of wall six feet long, made at the G.W.R. Concrete Depot at Taunton. These sections are comparatively light in weight (27 cwts.) and are keyed together so as to form a continuous wall of ample strength for requirements. The construction is not only economical, but it is also very rapid.

A similar but rather more ingenious innovation is the concrete cribwork used at Cardiff recently to form a retaining wall. The sections in this design also were made at the G.W.R. Concrete Depôt, and the interlocking effect seen in this illustration gives the cribwork strength as a retaining wall equal to what would be obtained by a much greater amount of brickwork or solid concrete. This also is a cheap and rapid system of construction.

Concrete
Facing
for
Platform
Extensions
at
Paddington

Concrete Walling at Cardiff

Reinforced concrete is being used in the various forms now in vogue for bridges, gantries, jetties, foundations, warehouses and other buildings. It is not yet known what the ultimate life of these structures will be, but with modern design and care in construction it is anticipated that a satisfactory life will be obtained and that a definite saving in maintenance costs will result.

The subject of the construction of railway buildings is far too wide for us to consider to-day, but you will, I think, be interested in the making of one Great Western Railway station where the circumstances were certainly out of the ordinary.

I refer to Fishguard Harbour station in West Wales. When construction was commenced the railway was seven miles distant at Letterston, and at the site selected precipitous cliffs of hard quartzite rock up to 200 feet in height ran sheer down to the sea, and afforded no ledge on which to begin operations.

The first job of all was to lower workmen by ropes from the cliff top in order to blast away enough rock to give a ledge on which to make a start on the work, for there was not even a foothold in the rock face. At the outset, cranes, engines, drills and all the paraphernalia required had to be brought by road from Letterston and lowered piece by piece from the cliff top.

Practically a whole hillside had to be removed by blasting in order to obtain a site for the station and harbour works, and about two million tons of rock were displaced to do so. As much as 113,000 tons were actually thrown down and masses of rock weighing as much as 5,000 tons were dislodged.

Year after year the work went on, many hundreds of workmen being continuously employed, until there grew up on the site an extensive quay, breakwater, a complete railway station with waiting rooms, refreshment rooms and platforms, over one-eighth of a mile in length, connected by subways and electric traversers ; running lines, sidings, signal boxes, offices ; a generating station for providing electric current for numerous cranes, lifts and other appliances ; in fact, an up-to-date and fully equipped railway station, seaport, and harbour.

∽　　∽　　∽　　∽

A large amount of station reconstruction has been undertaken in recent years, much of it in connection with the Government scheme introduced during the depression for the relief of unemployment.

You have seen, first hand, the wonderful improvements and additions which have been made at Paddington Station, and perhaps a talk about the reconstruction of Bristol Station would be of interest.

If you look at a railway map you will see that Bristol is a key point on the Great Western Railway, for it has lines radiating to South Wales, the Midlands, London, and the

Blasting the cliffs at Fishguard for station site

Fishguard Harbour Station

West of England, and it is, perhaps, excepting only Paddington itself, the most important station on the Great Western Railway.

Reconstruction of this station, which it is hoped will be completed during the present year, will result in the provision of fifteen platforms instead of nine ; the longest being 1,340 feet as against 920 feet previously, and the new station will cover about twice the area of the old.

The new building will be without overbridges, as all connections between platforms will be by means of a subway 300 feet long by 30 feet wide, with wide stairways for passengers, and electric lifts for luggage. The beautiful Gothic façade of the station will remain, but in the main entrance a spacious booking office with circulating area has been constructed, and on the other side of the station passimeter booking office, hairdressing saloons, baths, telegraph office, etc., have been provided. All the main platforms on the new station will be provided with refreshment and waiting rooms of modern design.

A feature of the new station is the provision of a new two-storey parcels depot, from which electric trolleys carry parcels traffic via lifts and a special parcels subway to all platforms.

The reconstruction work at Bristol has been carried on without interference with the heavy traffic, although the difficulties have been considerably increased by the station being situated on a curve between the River Avon Newcut and the Floating Harbour. The work has involved the reconstruction and widening of the river bridges, and no fewer than eleven road bridges have been widened which, incidentally, has necessitated the rebuilding of four local stations.

In addition to the station reconstruction extensive alterations have been made to the track, and the running lines for six and a-half miles in the station vicinity have been quadrupled. Additional running lines through the station will enable three trains to arrive and depart at the same time.

In permitting more and longer trains to be dealt with, the improvements will much facilitate the working through Bristol of trains to and from all parts of the country.

∽   ∽   ∽   ∽

Here we are back again at Bristol, considering a brand new station which is being provided for the city in which the Great Western Railway was born, where its famous first engineer was appointed, and where the original Great Western Railway terminated. This seems an appropriate place to bring this series of talks to a close, with the exception of a few concluding remarks, and these shall form the subject of our twentieth and final talk.

The striking façade at Bristol, Temple Meads Station

One of the platforms of the New Station at Bristol, Temple Meads

TALK NUMBER TWENTY

CONCLUSION

IN this series of railway talks we have discussed quite a number of " Track Topics." There are still many others which have to be omitted. In fact, the difficulty has been one of what to omit rather than of what to include.

As I have said before, our subject is extensive, and one realises its dimensions in an endeavour to select the more interesting aspects for a series of talks such as this. It has been a problem of compression as well as of selection, for any one of the subjects chosen would probably provide enough material for a whole series of talks if dealt with in any way exhaustively.

Before we part I should like you to appreciate the extent of the Great Western Railway system, in order the better to visualise the vast organisation which is necessary to keep the track and all that appertains to it in first-class order.

As you have been told, the G.W.R. consists largely of the amalgamation of a number of independent railways and it has been the policy since the early years of this century to straighten out the lines to make them conform, wherever commercially possible, to Euclid's definition of a straight line—" the shortest distance between two points."

It is, perhaps, not generally known that in this way the Great Western Railway has constructed nearly 200 miles

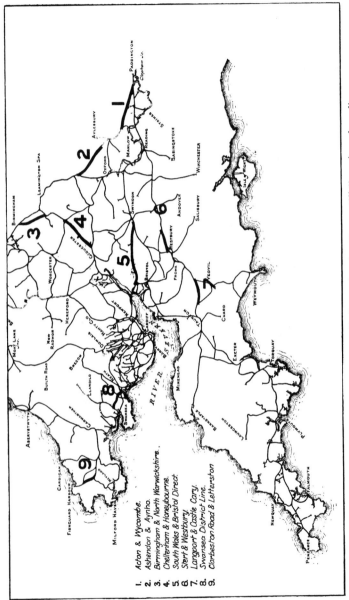

Diagram Map of Great Western Railway shewing "cut-off" lines (numbered)

1. *Acton & Wycombe.*
2. *Ashendon & Aynho.*
3. *Birmingham & North Warwickshire.*
4. *Cheltenham & Honeybourne.*
5. *South Wales & Bristol Direct.*
6. *Stert & Westbury.*
7. *Langport & Castle Cary.*
8. *Swansea District Line.*
9. *Clarbeston Road & Letterston.*

CONCLUSION

of new main line, at an expenditure of many millions of pounds, and relieved the undertaking of the charge made in other days of being the " *Great Way Round.*"

If you look at a map of the Great Western Railway you will see that its track serves (besides a thousand miles of ocean coast) a large part of industrial England and Wales as well as the West Country. It embraces, besides a number of lines only less important, three principal main lines.

1. LONDON TO PENZANCE.

   This route was formerly via Bristol. The line was shortened and improved by adapting existing lines and constructing link-up lines, reducing the distance—London to Penzance—from 326 miles via Bristol to 306 miles via Westbury.

2. LONDON TO WEST WALES.

   This route was much shortened (as we have seen) by the Severn Tunnel, and the mileage was further reduced by the provision of the South Wales and Bristol direct line, from Wootton Bassett to Filton, enabling services to Bristol (via Badminton) and Wales to be accelerated and improved.

3. LONDON TO BIRMINGHAM AND BIRKENHEAD.

   This route was formerly via Reading and Oxford. It was considerably shortened by the Acton and Northolt (G.W. & G.C. Joint Line) and the Ashendon and Aynho Line. In addition to affording improved access to north and west Welsh coast resorts, this line created the shortest route between London and Birmingham.

Concurrently with these developments new and shorter cross-country lines were constructed, improving the routes between Bristol and South Wales and Bristol and Birmingham and the North.

We hear, in these days of congested highways, a good deal about the construction of by-pass roads, but such by-passes have been a feature of railway construction for some considerable time, particularly on the Great Western Railway, and quite recently two by-passes have been provided on the route to the West of England, to cut off the towns of Westbury and Frome.

There is another cut-off which has some historic interest, and that is the west curve at Slough. This line is said to have been provided for the convenience of the late Queen Victoria, so that royal trains from Windsor to points west of Slough would have no need to enter that station.

ᔐ ᔐ ᔐ ᔐ

That arch-philosopher, Mr. Dooley, has been credited with the profound utterance "*if Nature had intended people to travel she would have put places nearer together.*" Well, all I can say is—and in this I am sure you will bear me out —if Mr. Dooley is correct, the Great Western Railway has done much to rectify Nature's shortcomings in the territory which it serves, and a principal share in the good work of bringing " places nearer together " has been taken by the Engineering Department in providing and maintaining such an excellent track.

I think I am fairly correct in saying that ever since the earliest days the Great Western Railway track has been

CONCLUSION

described as the smoothest in the world. The traditions of its famous first engineer, who laid the original track, are carried on by the Engineering Department of the Company to-day, and full justification for this claim can be found by making a journey by any of the G.W.R. express trains, which often travel at over eighty miles an hour.

The phrase, " They ' Go Great Western ' because the going's good," is one which you have heard before, but perhaps with the knowledge you have acquired about the activities of the Engineering Department, you now know what *makes* the going so good.

AU REVOIR

Above: Looking towards the 'Country' end at Paddington station.
Opposite: The shape of modern track maintenance, a high-output
ballast cleaner built by Plasser & Theurer. *(Network Rail)*

Track Topics Revisited

It is around eighty years since *Track Topics* was published in 1935 and this update section provides the opportunity to revisit some of the subjects covered and to assess how they have fared with the passage of time, as well as looking at the changing methods employed by the railway engineers.

For the most part it is good news as much of the GWR main line has survived intact through the vagaries of the Second World War and the subsequent post-war nationalisation of Britain's railways. It is all thanks to the man in the stove-pipe hat. Isambard Kingdom Brunel did such an outstanding job in building the Great Western Railway that it has, as they say, proved fit-for-purpose to this day. When Brunel surveyed and built the line from London to Bristol he was determined to make it the finest railway in the world. He achieved this by keeping the gradients to a minimum and by building on a grand scale to accommodate his broad gauge. Alas the broad gauge was defeated by its narrower-minded rival but his railway line has been little altered, apart from a doubling up of the tracks in

places to accommodate increases in rail traffic. It was not by chance that the London Bristol line was the first one chosen for the new HST 125s when they came into service in 1976.

Of the many examples of Brunel's civil engineering and architectural prowess the vast majority are with us to this day. Those which featured in *Track Topics* include the brick-built spans of the Wharncliffe Viaduct and the wide arches of the bridge at Maidenhead, plus Sonning Cutting, the Box and Middle Hill tunnels, the South Devon line on the sea wall and the magnificent Royal Albert Bridge over the River Tamar at Saltash. Many of the minor stations had already been lost in the widening of the main line, but Brunel's Temple Meads and Paddington are still with us, although of these only Paddington is used by the trains, for now at least. However there have been some notable casualties along the way. The broad gauge itself, of course, was condemned in Brunel's lifetime and finally ripped up in 1892, and then there are the numerous wooden viaducts which had all been replaced by the time *Track Topics* was published. Perhaps the greatest loss has been the Chepstow tubular railway bridge which was replaced in the 1960s as it could no longer cope with the heavy modern trains – see page 280.

Above: The Royal Albert Bridge during the recent refurbishment, photographed in 2012 from the Saltash side.
Opposite: Repairs being carried out on points. *(Network Rail)*

The greater challenge to the railways in the last eighty years has through the meteoric rise and dominance of road transport, which in the decades since the Second World War has seen the vast railway goods network stripped to its bare bones. Ironically it is the rise of the motor car, or at least the collateral impact of traffic congestion and pollution, that has brought travelers back to the railways in unprecedented numbers. More by neglect than design many of the major railway works, including some of the great London stations, have survived the economical doldrums and the drive for modernisation to be restored as the jewels in the modern transportation system. We may have lost Euston along the way but St Pancras, King's Cross and even Brunel's Paddington shine out, re-invented and re-invigorated in a way unimaginable as little as twenty or thirty years ago. The railways, the greatest of Victorian inventions, have turned out to be their bequest to the twenty-first century. Despite the many challenges still to come, not least the visual impact of the electrification of the old GWR main line, we have good cause to be optimistic for the survival of the engineering works featured in these pages.

The Wharncliffe Viaduct

One of first obstacles Brunel had to overcome in building the GWR was the Brent Valley on the western edge of London. *Track Topics* described this as 'quite the most formidable single engineering proposition in making the railway between London and the crossing of the Thames at Maidenhead'. The Wharncliffe Viaduct is massive at 896 feet long, but it is arguably one of Brunel's least appreciated works as it lacks the setting of a wide river as at Saltash, Chepstow or Clifton. The viaduct now strides a park which is mostly frequented by local dog walkers and at this point the Brent is almost lost in the undergrowth. It is well worth the trouble to go and see it, and when you get there note the coat-of-arms high up on the south-facing brickwork, put there by Brunel in acknowledgment to Lord Wharncliffe who had been chairman of the House of Lords committee which approved the building of the railway. The viaduct was widened in the 1890s, going from two broad gauge lines to four of the so-called standard gauge, and standing beneath the arches you can see the join where the new section and additional set of piers were added on the north side. Take a look at the brickwork, eight deep to form the arches, and also the piers, especially the way they taper and are capped in sandstone, and you will recognise the Egyptian styling repeated from the piers of the Clifton Suspension Bridge.

There are several other notable brick structures on this stretch of the line through the Thames valley. The more famous is the bridge over the river at Maidenhead which features two elliptical arches, the widest ever built, each with a span of 128 ft, and to the west of Reading there are the less well known skew bridges at Basildon and Mouslford; marvelous examples of brickwork on a compound curve. These are only accessible on foot.

• See pages 44-50.

Opposite: Two views of the Wharncliffe Viaduct which straddles the Brent Valley to the west of Hanwell. Consisting of eight semi-elliptical brick spans the original had just two piers, with the third added on the northern side when the track and viaduct were widened.

Tunnels

Tunnels are the great unseen wonders of the railways. Opened to traffic in June 1840, Box Tunnel between Chippenham and Bath is famed for its great length, 9,636 feet or 1.83 miles, and the cost in human lives with over 100 men killed during it construction. Unusually the tunnel descends on a 1:100 gradient from east to west. The ornate western portal is the most photographed, while the eastern one is very plain and sunk within a long and inaccessible cutting. The view of the western portal has recently been improved by clearing some of the obstructing foliage. Much-repeated stories that the rising sun shines from one end of the tunnel to the other on the morning of Brunel's birthday, 6 April, are unlikely to be true. The tunnel is straight enough but the position of the sun is not consistent due to variations caused by leap years. For my money the nearby Middle Hill Tunnel, although much shorter, has a finer portico but it is not easily seen close up.

• See pages 56-67, plus the Severn Tunnel 176-203.

Opposite: The western portal at Box, the longest tunnel on the GWR main line. The lower picture shows one of the six ventilation shafts.
Below: Another western portal, this time on the nearby Middle Hill Tunnel to the west of Box.

A section of atmospheric pipe on display at Didcot. A piston was sucked along the pipe, connected to the carriages through the slot at the top. *Below:* The truncated chimney of the pumping station at Starcross, and the old signal box on Dawlish station.

Atmospherics in South Devon

There can be few scenic routes to rival the South Devon line as it heads southwards from Exeter St Davids to skirt along the bank of the Exe estuary, then hugs the coast through Dawlish and Teignmouth before heading west and inland to Newton Abbot. This line was the location of Brunel's ill-fated atmospheric experiment which is described in some detail in *Track Topics*. Three of the pumping stations which housed the steam engines are still standing. At Starcross the pump-house, built of deep red sandstone, is sandwiched between the road and the railway with its truncated chimney looking more like a church tower nowadays. The most complete example is at Torquay, although this was never used by the time that the atmospheric scheme was abandoned in September 1848. It can be viewed from the car park of the Longpark supermarket. Further down the line the third survivor, at Totnes, has become part of a milk processing plant located immediately next to the station. Partial remains of other pump-houses can also be seen near Turf Lock and also at the rear of the station car park at Dawlish. Sections of the atmospheric pipe are displayed at various museums with one of the longest examples at the Didcot Railway Centre.

The trains still use the coastal line which runs along the sea wall nestling at the foot of the distinctive red cliffs before cutting across the front of Dawlish separating the town from the beach. This area wasn't chosen for its picturesque qualities, it was simply the best route to avoid the hilly terrain further inland. You can't imagine that the idea of building a railway along such a beautiful piece of coastline would get past today's planning committees, but here it is and this stretch of line continuing as far as Teignmouth is much favoured by railway photographers.

• See pages 70-87.

The Royal Albert Bridge

By 2013 work was nearing completion on a £10 million refurbishment of the Royal Albert Bridge which crosses the River Tamar to link Devon and Cornwall. It is a remarkable tribute to Brunel that such care is lavished on a bridge which is over 150 years old and, after all, only carries a single track. Work started in 2010 and the engineers were expected to spend nearly two million hours of work strengthening and repainting the bridge. The latter process has thrown up an interesting discovery. By stripping away countless layers of paint the engineers have revealed that when completed in 1859 the spans were finished in a pale stone or off-white colour. That didn't last long though, and red-brown anti-rust paint was applied to the main spans within a decade, along with the goose-gray which we are more familiar with today. The current refurb will restore the bridge to the gray which was the colour when it was Grade 1 listed in 1952.

As *Track Topics* points out previous improvements have been made to the bridge, with 401 new cross girders added to increase stiffness along with the reconstruction of the first two spans on the Cornwall side in 1905. Then again in 1928 when it was decided to replace the girders on the remaining land spans. The current refurbishment has been carried out concurrently from each end of the bridge and to provide a safe working environment, as well to contain any dust and debris from falling into the river, the sections being worked on have been encapsulated in tent-like structures.

The most obvious new addition at Saltash is the A38 Tamar Road Bridge which opened in 1961. Admittedly this obscures the view of Brunel's bridge from the northern side, but it does provide a perfect vantage point for pedestrians. Note that the road bridge has a toll for cars going into Devon, although none for cyclists or pedestrians. If you park on the Devon side and walk into Saltash, look out for the bust of Brunel where North Road crosses the A38. It also worth going down to the riverside for a fabulous view of the bridge, but try to ignore the newer, bright green statue of IKB.

• See pages 88-100.

Opposite: Facing towards Saltash, this photograph was taken from the road bridge in 2012 during the recent refurbishment. *Below:* Looking along the undulating track on the Royal Albert Bridge. *(Network Rail)*

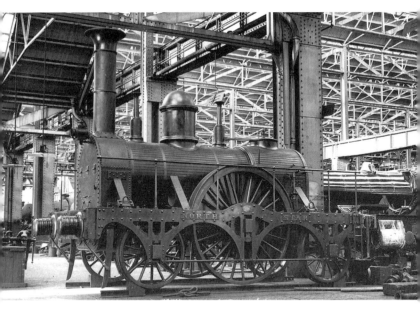

Replica *North Star* inside the Swindon Works, probably photographed around the time of the 1935 bicentenary, and *Iron Duke* standing in the open at Toddington, Gloucestershire.

Broad Gauge Relics

The broad gauge was Brunel's great experiment to create a better railway. Condemned by parliamentary commission in 1846, and abolished in 1892, it was also his greatest failure. Apart from the line itself, including the stations and several fine examples of goods sheds such as the one at Stroud in Gloucestershire, very little remains of the broad gauge trains themselves.

North Star, originally built by Stephenson for an American client and converted to the GWR's 7 ft rails, was one of the first locomotives to run on the GWR and certainly the first successful one. Withdrawn in 1871, it survived the great cull of broad gauge locomotives in the 1890s and was preserved at Swindon as a relic of the early days. However it was scrapped in 1906 to make more room in the workshops, with only the huge 7 ft diameter driving wheels remaining. Perversely the full-size replica, now displayed at the GWR Steam Museum in Swindon, was created to mark the 100th anniversary of the Stockton & Darlington Railway in 1925 and wheeled out again for the cameras for the GWR's own centenary in 1935. (The year in which *Track Topics* was published.) This non-working replica incorporates the original driving wheels.

There are two other broad gauge replicas of note. In 1985 a working replica of Daniel Gooch's massive *Iron Duke* was created using donar parts from two Hunslet Austerity shunters. Built to take part in the GWR 150 celebrations, it is longer capable of being steamed and is currently displayed at the Gloucestershire Warwickshire Railway's Toddington station in Gloucester. The third replica is the *Fire Fly*, an earlier Gooch design, completed in 2005 in time for Brunel's bicentenary in 2006. This runs in steam on a regular basis together pulling replica coaches on a stretch of broad gauge track at the Didcot Railway Centre. Clearly we will have to wait for the next big anniversary to see another replica. The only surviving original broad gauge loco is *Tiny*, a diminutive vertical-tank 0-4-0 shunter built for dock work. No longer running, this is displayed by the South Devon Railway at Buckfastleigh.

- See pages 26-43.

On Today's Tracks

A big chunk of *Track Topics* is devoted to the subject of laying and maintaining the permanent way – the ballast and rails – and this is the one area of railway activity that has seen the greatest changes since the 1930s. Gone are the teams of trackmen using brute strength and lining bars to correct track alignment, or shovels and picks to pack and tamp the ballast. New methods in track laying have resulted in a smoother and faster ride, and the mechanisation of these processes has been vital in minimising the disruption to travellers.

Traditionally track laying was carried out using fixed lengths of hot-rolled steel rail – usually 66 ft long in the UK – which was joined end to end using bolts and fishplates. The familiar 'clackety-clack' sound was made when the wheels of the train passed over the small gaps between the jointed rails. Not only did this construction method demand a high level of attention to maintain the geometry of the rails and the smooth running of the trains, it wasn't suited to the modern heavier and faster trains. There was also a major issue with cracking around the bolt holes on the fishplates which could lead to breaking of the running surface. This was the cause of the Hither Green rail crash in November 1967, an event which accelerated the conversion of much of the UK network to continuous welded rail. As a result Network Rail now deploys an army of incredible track laying machines which work through the night to renew the permanent way and make the clackety-clack a thing of the past on all major lines.

The biggest of the machines in Network Rail's stable are the two high-output track laying trains built by the Austrian specialist rail equipment makers Plasser & Theurer. Known by Network Rail as Track Renewal Systems TRS2 and TRS4, each train is more than 2,600 ft long and weighs in at 2,500 tons. It has been claimed that

Opposite: The big TRS4 – Track Renewal System – in action on the West Coast Main Line in 2012. Major work on the track is usually carried out at night or at weekends to minimise the disruption to train services. *(Network Rail)*

they lay new track faster than you can walk, and during recent work on renewing the West Coast Main Line TRS4 was putting down track at a rate of over 2,100 ft in an eight hour overnight session, or almost three time that much during a sixteen-hour weekend line closure. Working alongside a high-output ballast cleaner, the track laying system operates on a continuous process. Firstly, it unclips the old sleeper fastenings and removes them from the track with a magnetic drum, and then lifts the rail from the sleeper housings. The old sleepers are lifted and transported to special sleeper wagons using gantry cranes and pallet style containers. It then levels off the ballast bed and displaces the material to the side of the track. New sleepers are placed on the prepared ballast bed and spaced correctly, after which new rail is placed in position and fastened to the sleepers. The rails are flash butt welded by the application of a strong electrical current through the touching ends of two sections of rail. The ends become white hot and are pressed together to form a strong weld.

Finally a second train cleans and collects the ballast and distributes it back to the newly installed track, ready for final geometry correction with a tramping machine.

Switches and crossings are factory-built as a series of panels and these are delivered to site on special wagons that tilt sideways so that they can fit within the restrictive width of the railway lines. An integrated lifting and handling system eliminates the need to bring large cranes on site or onto land adjacent to the railway.

Ballast-less track is an option to the traditional sleeper and ballast permanent way, laid either as a continuous reinforced concrete slab or in precast units, but this is more expensive to lay and causes far longer closure times for the railway operators. It is not favoured in the UK and the present system of high-output track renewal used by Network Rail has seen closure times reduced from days to hours. This means that the trains are kept running for longer which is better for everyone.

• See pages 117-136, 137-153, 154-162, 163-175, 213-218.

Opposite: The latest mechanised track laying system, TRS4, at work on the West Coast Main Line. *Above:* Mechanical slinger moving concrete sleepers. *(Network Rail)*

Viaducts

Apart from the conventional brick and masonry viaducts throughout the GWR network, little mention is made of the timber viaducts designed by Brunel. Closely associated with the many Cornish valleys, they were also built in a number of other locations in the south west, the West Midlands and also in South Wales. (As you will see on pages 64 and 281 IKB was not averse to using timber on some of his bridges, such as skew bridge at Bath.) In the main he used timber in order to keep costs down. This was at a time when Baltic pine was readily available and relatively cheap, and once treated with preservatives the timber elements of these structures were expected to last many years before being replaced with like for like. However by the time *Track Topics* was published the wooden viaducts were seen as unwanted relics of the past and their replacement was a natural part of the continuing modernisation of the railways. The only examples depicted in the book were either undergoing replacement, or had already been replaced. Most of the Cornish timber viaducts had gone by the end the nineteenth century, and the last survivors on the quieter Falmouth branch line were replaced by 1934.

In some cases where the original masonry piers were strong enough the upper wooden sections were given masonry or brick extensions and the rails carried on new steel supports and decking. If the piers were not substantial enough, or sometimes because they were entirely constructed of timber such as the one at St Germans, complete new brick viaducts were built alongside the originals. There are several locations were the old unused masonry towers have been left standing like rows of broken teeth. Examples are to be found throughout Cornwall including those at Moorswater, St Austell and Truro, all on the main line, and also at Ponsanooth and Carnon on the Falmouth branch.

Opposite: Coldrennick Viaduct immediately east of Menheniot. The timber structure was removed in 1898 and replaced by brick extensions and a steel deck. *Inset: Track Topics* frontispiece image showing rebuilding of the Ponsanooth Viaduct on the Falmouth branch line.

Top: Chepstow railway bridge, or at least its modern replacement featuring the original iron columns. *Left:* A section of the iron bridge deck at Brunel University, Uxbridge. *Above:* In 1962 the superstructure of the old bridge was dismantled one side at a time.

Bridges

Apart from the Royal Albert Bridge at Saltash, there is mixed news on the bridges. Inevitably these take the strain of the modern heavy trains and increased rail traffic, and many of Brunel's lesser-known railway bridges have gone to make way for wider or stronger structures. The masonry bridge over the Avon leading into Temple Meads, for example, is still there but is now obscured by later girder bridges added to either side. However the most significant loss since the publication of *Track Topics* is undoubtedly the distinctive tubular bridge which crossed the River Wye at Chepstow on the border between England and Wales. Historically this was an important bridge as its wrought-iron tubular construction led directly to the design of the later Saltash bridge, but by the 1960s the old structure had become too weak and a pragmatic and unsentimental British Rail replaced it with the present under-slung steel girder truss in 1962. The original iron support columns still remain. In 1988 the A48 road bridge was added alongside the railway bridge.

• See 219-235, Royal Albert Bridge 88-100 and update 270-271.

Below: Immediately to the west of Bath station the line crosses over the Avon once again. This is the modern replacement for Brunel's wooden skew bridge. See page 65.

Brunel's frontage for the Temple Meads terminus in an Elizabethan style. Passengers arriving by horse-drawn carriages entered via the archway on the left-hand side, passed under the tracks and emerge on the right-hand side to depart through a complementary arch, later demolished to improve the approach road. The square block rising above the roof is a water tower.

Middle left: The extension to the Brunel passenger shed loosely repeats its profile and part of the old roof and three of the mock hammer-beams have been left stranded on this side of the more recent dividing wall. Note the wall-mounted signal box on the right.

Bottom: The clock tower and main entrance to the 1870s extension to Temple Meads.

Bristol Temple Meads

The world's first purpose-built railway terminus was designed by Brunel with an Elizabethan facade and housed the booking offices and meeting room of the Bristol board of the GWR, with the engine and passenger sheds behind. Located on the left-hand side of the approach road these should not be confused with the later non-Brunellian extension which forms the major part of the present station. (Note from the picture on page 253 that the clock tower lost its characteristically late-Victorian roof during the Second World War.) The engine shed immediately behind the offices was occupied by a museum temporarily, and Brunel's passenger shed occasionally serves as an event venue although it is not otherwise generally accessible to the public. The extension is now a car park, for the moment at least, with the rails cut off by new buildings. Note the slender wall-piercing signal-box. In the 1930s Temple Meads was further extended with several new platforms added.

• See pages 246-255.

Platforms 10 and 12 at Temple Meads showing the more modernistic white tiles and styling of the 1930s extension.

Paddington Station

The London terminus for the GWR is in great shape. It has avoided the development creep that has blighted so many of the capital's high-profile stations, and in the 1990s Brunel's magnificent roof of wrought-iron and glass, arranged as three 700 ft-long transepts or spans, underwent a major refurbishment. The paintwork was restored to its original colour and in a weight-saving exercise the corrugated iron cladding was replaced with profiled metal sheeting and the glass by polycarbonate glazing. The long platforms have been cleared of years of accumulated clutter, including many cumbersome information boards, and there is new limestone flooring throughout. At the town end the 'Lawn' area has become an oasis of calm where the cappuccino-sippers can stroke their smartphones in peace caged behind a 150 ft-long glass screen. These improvements have restored unimpeded views of the curved ribs of the roof, leading all the way to the delicate tracery of the glass screen at the Country end.

In part this multi-million pound facelift was the sugar coating on a bitter pill. The second phase of the 'reduction and redevelopment' process was to demolish the fourth non-Brunellian span on the eastern side of the station. Added in the early twentieth century, the developers argued that although it loosely echoed Brunel's spans – albeit slightly bigger and constructed of sturdier steel not wrought iron – its removal was necessary to make way for a new Crossrail station. Its loss would return Paddington to its original, purest state. Hogwash said the many objectors to the scheme. The fourth span had been there long enough to become an integral part of Paddington, and its curving backbone had greeted generations of travellers descending the slope from Praed Street. Thankfully the campaigners got their way and the Crossrail station is now nearing completion out of sight underneath Eastbourne Terrace on the western side. Span 4 has also been restored.

• See pages 246-255.

Opposite top: Span 3 and HST in 2012, with Span 4 to the right.
Bottom: The view from Bishop's Bridge Road with Brunel's three spans on the right and the later Edwardian-era fourth span on the left.

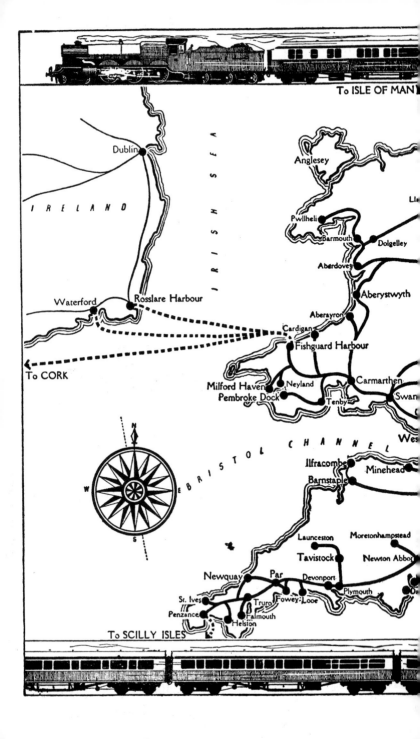